DPs Dominion
Publishing Services
http://www.dominionpublishingstores.yolasite.com

Timeline of Agricultural Mechanization

The evolution, the revolution, the revolutionists

Segun R. Bello
B. Eng (Hons), FUT, Akure, MSc, Ibadan,
MNSE, MNIAE, FSINRHD, R. Engr. (COREN)

Timeline of Agricultural Mechanization
The evolution, the revolution, the revolutionists

Copyright © 2014 Bello, RS
Federal College of Agriculture Ishiagu,
480001 Nigeria
segemi2002@gmail.com; segemi2002@yahoo.com; bellraph95@yahoo.com
http://www.segzybrap.web.com
+234 8068576763, +234 8062432694

All Rights Reserved

No part of this book may be reproduced, stored in a retrieval system or transmitted, in any form or by any means, electronics, mechanical, photocopying, recording or otherwise, without the prior written permission of the copyright holder.

ISBN-13: 978-149-498-361-1
ISBN-   149-498-361-3

First published March 2014
Printed by Createspace US

Timeline of agricultural mechanization

Timeline Quote

"I have been branded with folly and madness for attempting what the world calls impossibilities, and even from the great engineer, the late Mr. James Watt, who said to an eminent scientific character still living [in my time], that I deserved hanging for bringing into use the high-pressure engine. This so far has been my reward from the public; but should this be all, I shall be satisfied by the great secret pleasure and laudable pride that I feel in my own breast from having been the instrument of bringing forward and nurturing new principles and new arrangements of boundless value to my country. However much I may be straitened in pecuniary circumstances, the great honour of being a useful subject can never be taken from me, which to me far exceeds riches".

♣ *Richard Trevithick (1771-1833)*

Dedication

To the glory of God Almighty

Acknowledgements

I sincerely appreciate all present and past students of the department of Agricultural and Bio-environmental Engineering Technology, Federal College of Agriculture, Ishiagu and the numerous individuals whose reports indicated the impact my contributions had made in solving some of their academic challenges. The contributions of my professional colleagues and friends during the over fifteen years of teaching and research as an agricultural engineer and the wealth of experiences thus far gathered in the course of my career are immeasurable and I appreciate them.

My special thanks go to my friend and dear wife, who had always backed-up the realization of God's plan for me. She is truly virtuous and a help-meet indeed. I am grateful for her understanding and tolerance in taking full responsibility of running our home during the preparation and review of the manuscript.

I am equally grateful to my lovely children, Ayomikun, Pelumi Damilola and Adeola, who are epitome of love and care. I am encouraged and strengthened by their prayers, may God bless all.

Jesus is Lord.

Timeline of agricultural mechanization

Content

Dedication ix
Acknowledgements xi
Preface xvii

Part 1 HISTORIC HUNTER-GATHERING TRANSITION: THE EVOLUTION 1

Chapter 1   AGRICULTURE: The Historic Developments 3
1   Introduction 3
1.1   The age of man 4
1.2   Man in the pre-agricultural times 5
1.3   Agriculture 6
1.3.1   Origin of agriculture 7
1.3.2   History of agriculture 8
1.4   Early agricultural sites in the world 10
1.5   Plant and animal domestication 15
1.6   Agriculture: A revolution or an evolution? 16
1.7   Agricultural evolution: History and perspective 17
1.8   Agricultural revolution: History and perspective 18
1.9   Phases of agricultural development 25
1.10   Agricultural revolution: History and perspectives 32
1.11   The green revolution 36
1.12   History of agricultural development in Nigeria 37

CHAPTER 2   AGRICULTURAL SOCIETY: The Historic Developments 47
2.0   Society in the middle ages 47
2.1   Agricultural Society 48
2.2   Characteristics of agricultural societies 48
2.3   Features of early agricultural societies 49
2.3.1   Hunter-gatherer society 49
2.3.2   The agrarian society 52
2.3.3   Agricultural revolution society 52

CHAPTER 3   ENGINEERING SOCIETY: The Historic Developments 55
3.0   Introduction 55
3.1   Engineering society 55
3.2   Status of engineers in society 55
3.3   History of agriculture and engineering 56
3.4   Historic development of Agricultural Engineering Society 56

CHAPTER 4	AGRICULTURAL ACTIVITIES: The Historic Developments .. 60
4.0	Introduction 60
4.1	Agriculture in the biblical era 60
4.2	Farming in the early agricultural era 63
4.2.1	Traditional farming activities in the early agricultural era 64
4.2.2	Mechanized farming activities in the early agricultural era 70

Part 2 THE AGRARIAN MECHANIZATION: THE REVOLUTION 84

CHAPTER 5	MECHANIZATION: The Historic Developments 86
5.0	Introduction 86
5.1	Mechanized agriculture 87
5.2	Historic tool development in primitive agricultural era 91
5.3	Historic tools development in traditional agricultural era 97
5.4	Historic power development in mechanized agricultural era 98

CHAPTER 6	AGRICULTURAL TRACTORS: The Historic Developments 104
6	Introduction 104
6.1	Historic development of agricultural engines 104
6.2	Historic development of agricultural tractors 134
6.3	Historic innovations in tractor development 147
6.4	Future of tractor development 159
6.5	Most significant innovations in farm power development 161

CHAPTER 7	AGRICULTURAL MACHINERY: Historic Developments 163
7.0	Introduction 163
7.1	History of land clearing and development 164
7.2	History of farm machinery development 166
7.3	Historic developments in crop planting and seed sowing 197
7.4	Historic development in crop harvest and processing machinery 212
7.5	History of animal products processing 253
7.6	History of farm transport development 255

Part 3 THE AGRARIAN REVOLUTIONISTS: HISTORY & ENTERPRISE 266

CHAPTER 8	AGRARIAN REVOLUTIONISTS: Their History 268
8.0	Introduction 268
8.1	The agrarian revolutionists: their lives and inventions 268
8.2	The agrarian revolutionists: their enterprises and growth 292
References	 303
Appendix	 316

Timeline of agricultural mechanization

Preface

Agriculture is without doubt one of the greatest revolutionary innovations that started in the 19th century as evident in the 21st century. In order to sustainably manage the agro-ecological systems created by such innovations in the face of growing and potentially more prosperous global population, the historical antecedents upon which the transformation and the revolutionary development that brought about the present time and what the future of the world agriculture holds become imperative.

The demands for food supply, socio-economic factors, environmental and technological revolution as well as political factors; many of which are spatially specific, have been identified as tools that shaped the characteristic trends in agricultural systems evolution. It is important to know that the performance of agriculture measured (before now) in terms of food, fiber and bio-energy production is now, to a large extent, measured on a range of other social and environmental outcomes, positive and negative. However, the emerging consensus of agricultural mechanization at meeting the needs of a growing and potentially more prosperous global population mainly from the existing stock of agricultural resources has gone a long way in saving the world's ecosystems that could have been irreparably damaged.

These needs therefore necessitated a study of the historic events that led to agricultural evolution, revolution, and developments that transcends the old and the new worlds to the 18th century agricultural and industrial revolutions; the 19th century innovations in tools mechanization and power developments, the 20th century complex machines and equipment design and manufacturing among others, which this book gave an explicit account of from the pre-agricultural times (over 3million years ago) through the agricultural times (10,000 years ago) into the agro-based times (19th and 20th centuries) and the emerging sustainable technologies in agricultural mechanization of the 21st century, thereby incorporating future developments. The book is in three segments:

Part 1 x-ray the interrelationships between early hunter-gatherer society, the agricultural society, engineering society; and the history of man's creation and survival from the historic biblical records, his activities in the generic Garden of Eden, the fall, the departure, the commission (to till the ground) as well as animal hunting. The first society led the transformation (evolution) into the agrarian society, through the revolution into the machine automation agriculture.

Part 2 explored the historic developments in the agrarian (agricultural mechanization) era throughout all the ages of man's development until the last agricultural revolution which started in the 1950s. The timeline of agricultural mechanization inventions in America until late 20th century is also presented in part.

Part 3 made a review of the biographies of the most influential scientific inventors and revolutionaries who had changed the face of agricultural development from the Paleolithic age through the last Iron Age into the 3rd agrarian revolution. Their history, their inventions, their enterprise, lifestyles, failures and end were discussed in part 3.

This book adequately meets the requirement for academic and curriculum studies to earn you a degree. It is a must have, and a must read research documentation packaged to get us acquainted with historic developments in agriculture and agricultural mechanization. This scientific story book transcends events of four major active agricultural centuries and the activities and stories of notable inventors that help draw inspirations from their hard works, their determinations, their entrepreneurial dispositions, failures and above all their successes which had distinguished and gave them a place in history.

Bello RS,
480001, Nigeria

Timeline of agricultural mechanization

Part 1

HISTORIC HUNTER-GATHERING TRANSITION

THE EVOLUTION

Hunting and gathering in the Paleolithic era

Chapter 1

AGRICULTURE
The Historic developments

1   Introduction

The history of its evolution and the development had made agriculture one of the most innovative sectors in human history. However, agriculture had seldom been viewed as an innovative sector because the importance and contributions of agriculture to the overall economic growth and development has more often than not been ignored. Erroneously, what had been considered an innovation is mostly development in the input industries, not in the products alone, but also in the processes that evolved them. Nevertheless agricultural innovation and technological changes over the past centuries has globally improved food security and helped lower food prices. Agriculture did not just exist; it evolved, it grew over a period, and as a result has a history which qualifies it as an innovative sector, as will be seen in this book. How did agriculture evolved?

The very beginning of agricultural practice

The age of man, beyond scientific knowledge and archaeological postulations, dated back to God the Creator in the beginning. The primary purpose of man at creation was to exercise authority (dominion) over every other God's creation, living and non-living.

> *[So God created man in his own image... and God blessed them and said... be fruitful, and multiply and fill the earth; and have dominion... (Gen 1:27-30)]*

His first assignment was ultimately to tend the Garden planted (established) at the East of Eden.

[And the Lord planted a garden eastward in Eden; and there he put the man whom he had formed... and the Lord took the man, and put him into the garden of Eden to <u>till</u> [work] it and to keep it (Gen 2:8-15)]

When man erred he fell; he was driven out of the Garden of the Lord. After the fall of man, he was sent out of his comfort zone and banished to practice the first career bestowed on him by the virtue of God's judgment for disobedience; hunting and gathering (ultimately, the primitive agriculture as we know it today).

[Therefore the Lord God sent him (Adam) forth from the Garden of Eden, to till the ground from where he was taken (Gen 3:23)]

His immediate need then was survival and comfort; therefore his generation became hunters (wanderer), and hewer of wood and gatherer of water

[... and Abel (the second son of Adam) was a keeper of sheep, but Cain (the first born of Adam) was a tiller of ground (Gen 4:2)] [Emphasis mine]

These records undoubtly justify and characterized the primitive and pre-agricultural times spanning several centuries referenced by the timeline of the period before the birth of Jesus Christ [B.C.], and periods after His birth, [AD.], before the evolution of modern agricultural and agro-based times as practiced today (i.e. agronomy, horticulture, animal husbandry, irrigation and technology).

1.1   The age of man

Hesoid, a Greek poet who lived between the 8th-7th century B.C classified the Ages of Man as follows:

1. *The Golden Age* also known as the *Palaeolithic* Old Stone Age: The Palaeolithic age or better referred to as pre-agricultural societies began about 2.5million years ago. It lasted until about 8000 BC. Palaeolithic people were nomads who survived by hunting and gathering of foods. During the Palaeolithic Age, people discovered how to make fire, developed better tools and spoken language. The age was characterized by the age of the hunter-gatherer and the Eden-like pre-agricultural era.
2. *The Silver Age* (*The Ice Age* 8,000 B.C.): The ice age began about 8,000 BC during the period in which the level of water in the oceans went down. Humans survived the Ice Age by adapting to their surroundings. They changed their diets, built sturdier shelters, and used animal skins to make warmer clothing. During the ice age, a land bridge was revealed connecting the Asia and North America. This age was characterized by the birth of the concept of work and symbolized by the "Yoke of Oxen"

3. *The Neolithic Age*: The Neolithic age began around 8000 BC at the end of the last Ice age, and lasted until around 4000 BC. During the Neolithic Age, farming began to replace hunting and gathering. People also tamed animals. An increased food supply made it possible for people to settle in villages, take up jobs other than farming. The move toward settled farming gave rise to the agricultural revolution. At the end of Neolithic age, people discovered that mixing copper and tin formed bronze, which was stronger than copper alone. This discovery marked the beginning of the industrial revolution.
4. *Bronze Age* (3,500 B.C.): The time known as the Bronze Age lasted between 3500 and 1200 BC. Within this period, the use of bronze spread. Four great river valley civilizations emerged during the Bronze Age as follows- the Mesopotamia, Egypt, India, and China civilizations. Each of these early civilizations developed cities and formed governments. Struggle for dominance among the kingdoms led to bitter wars, the fall of old kingdoms and the rise of new ones at about 9th century BC, especially the Assyrians, who dominated all of Syria-Palestine, Egypt, and the Babylonia. However, this empire began to collapse toward the end of the 7th century BC, and was obliterated by an alliance between a resurgent New Kingdom of Babylonia and the Iranian Medes.
5. *Iron Age* (1,500-600 B.C.): The destruction at the end of the Bronze Age left a number of tiny kingdoms and City-states behind. During this period, a number of technological innovations spread, most notably iron working industries around the 16th century BC.

1.2 Man in the pre-agricultural times

Historical evidences and documentations on the existence of human species, known as Homo sapiens, were estimated to be 2,500,000 years old! For most of that time, during (the Adamic and the Old Stone Age era), humans survived as foragers or hunter-gatherers largely characterized by gathering of wild plants and hunting of animals. For most of their early history, humans survived this way in their natural environment. Foraging lifestyle is a subsistence form of lifestyle with short term goals and rewards.

Figure 1-1: Foraging in the early twentieth century with a beater

Figure 1-1 shows a Pomo[1] woman, Native North Americans Northwest of the Pacific, foraging in the early twentieth century; and using a beater to gather seeds into a basket. These early foragers knew which plants were edible, which were poisonous, and which had medicinal properties. They knew which plants could be used as dyes, which could be used for weaving or building material.). Figure 1-2 equally shows a Hupa[2] man hunting salmon in the early twentieth century with a spear.

Figure 1-2: Hunting in the early twentieth century with a spear, ca[3] 1923

These activities continued over centauries until the evolution of farm structural development otherwise known as agriculture. After the appearance of farming it was only gradually that they were absorbed into, or eliminated by, the encroachment of agriculture. A few hunter gatherer peoples survive to this day, but the world of the hunter gatherers, in which most ancient people followed this mode of life, is long gone, disappearing in the millennia following 10,000 BC.

1.3   Agriculture

Agriculture is a land-based primary industry that directly depends on natural resources such as land, water, and a diversity of plants and animals to supply man's food and fiber needs. It is supported by the application of human knowledge, both traditional and scientific, and human effort, skills and endeavour. Agriculture covers nearly 40% of the

---

[1] The Pomo people are an indigenous people of California in the USA. The historic Pomo territory in northern California was large, bordered by the Pacific Coast to the west extending inland to Clear Lake, and mainly between Cleone and Duncans Point.
[2] Hupa, also spelled Hoopa, is a Native American tribe in northwestern California. Their autonym is Natinixwe, also spelled Natinookwa, meaning "People of the place where the trails return".
[3] Ca or c. implies circa i.e. about

surface area of the earth and results in 30% of global greenhouse gas emissions and 70% of global water withdrawal (Sakrabani *et al.*, 2012).

1.3.1 Origin of agriculture

There have been several historic assertions and speculations that agriculture started about 10,000 years ago when the early foragers moved away from foraging to farming. Earliest archaeological evidences supported this historic assertion in several areas of the world. Evidence of the beginnings of agriculture have been reportedly found in the eastern North America, the Tehuacan and Oaxaca valleys in Mexico (called the Mesoamerica), the South American highlands, the Fertile Crescent of the Near East, the Yangtze and Yellow River valleys of China (Far East), and the New Guinea Highlands.

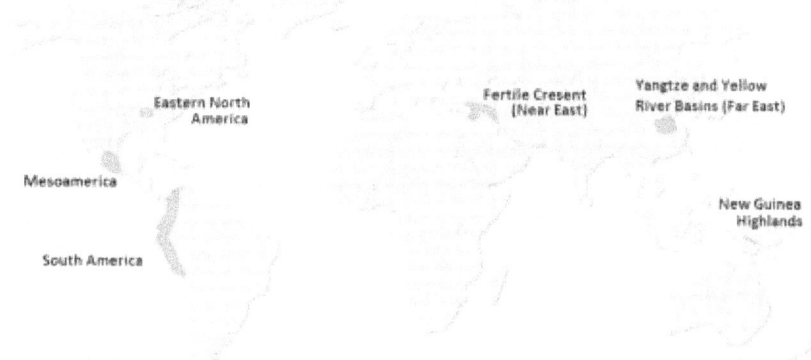

Figure 1-3: Evidences of the beginnings of agriculture

Research evidence also reported the appearance of agriculture in Europe in the modern-day Turkey around 8,500 years ago, spreading to France by about 7,800 years ago and then to Britain, Ireland and to Northern Europe approximately 6,000 years ago. However, the ancient Near East and the Fertile Crescent in particular, are generally seen as the birthplace of agriculture.

By this, over the following few thousand years thereafter, agriculture was reported to had flourished in the fertile crescent of the Near East, the plains of the Far East, and Mesoamerica. Agriculture came later to Europe, perhaps only 6,000 or 2,000 years ago. Great Britain was still making its transition to agriculture at the time Caesar's army entered around 100 BC. The paleo-anthropological evidence shows that with the advent of agricultural revolution of the 18th century, there was a decline in stature, cranial capacity, and muscularity, along with a general decline in human health and nutrition.

1.3.2 History of agriculture

The history of agriculture is closely tied to changes in climate, or so it certainly seems from the archaeological and environmental evidence. After the Last Glacial Maximum (LGM), what scholars call the last time the glacial ice was at its deepest and extended the farthest from the poles, the northern hemisphere of the planet began a slow warming trend (global warming phenomenon). The glaciers retreated back to the poles, and forested areas began to develop where tundra had been.

Figure 1-4: Timeline history of agriculture

This section takes a critical look at history of agricultural development from the origin through the evolutions, revolutions and or transformation that have taken place in agricultural operations, machinery and equipment development from the pre-agricultural times (over 3 million years ago) through the agricultural times (10,000 years ago) and agro-industrial times (19th century) and the intended 4th agricultural revolution era.

Pre-agricultural times

By the beginning of the late Epipaleolithic (or Mesolithic/Natufian) period, people had began to move into the newly opened areas northward, and lived in larger, more sedentary communities. The large-bodied mammals and humans which had survived on for thousands of years had disappeared, and now the people broadened their resource base, hunting smaller games such as gazelle, deer and rabbit, and gathering seeds from wild stands of wheat and barley, and collecting legumes and acorns. But, about 10,800 BC, the abrupt and brutal cold called the Younger Dryas (YD) by scholars occurred, and the glaciers returned to Europe, and the forested areas shrank or disappeared. The YD lasted for some 1,200 years, during which time people survived as best as they could.

Figure 1-5: Paleohunting of large-bodied mammals

Agricultural times

After the cold retracted, the climate rebounded quickly. People settled into large communities and developed complex social organizations, particularly in the Levant [4], where the *Natufian period* was established. Natufian people lived in year-round established communities and developed extensive trade systems to facilitate the movement of black basalt for ground stone tools, obsidian for chipped stone tools, and seashells for personal decoration. The first stone built structures were built in the Zagros[5] Mountains, where people collected seeds from wild cereals and captured wild sheep.

The PreCeramic Neolithic period saw the gradual intensification of the collection of wild cereals, and by 8,000 BC, fully domesticated versions of einkorn wheat, barley and chickpeas appeared, and sheep, goat, cattle and pig were in use within the hilly flanks of the Zagros Mountains, spreading outward from there over the next thousand years.

Scholars had debated the invention of farming, which was viewed as a labour-intensive way of living compared to hunting and gathering. Views suggested that domesticating animals and plants was seen as a more reliable food source than hunting and gathering. For whatever reason, by 8,000 BC, the die was cast, and humankind had turned towards agriculture.

Agro-industrial times

The agro-industrial times gave rise to massive and rapid increase in agricultural productivity and vast improvements in farm technology, which culminates in the agricultural revolution. A number of technological advances made the agro-industrial revolution possible led by a British inventor, Jethro Tull (1674-1741), widely regarded as the father of agrarian revolution.

In the 17th century seeds were sown by hand, but at about 1701, Tull had invented the seed drill and also invented a horse drawn hoe which killed weeds between rows of seeds. Furthermore new forms of crop rotation were introduced among several other agro-industrial activities.

---

[4] The Levant, also known as the Eastern Mediterranean, is a geographic and cultural region consisting of the "eastern Mediterranean littoral between Anatolia and Egypt
[5] The Zagros Mountains form the largest mountain range in the present day Iran and Iraq. **Zagros Mountain** range (about 550 miles (900 km) long and more than 150 miles (240 km) wide) in southwestern Iran, extended northwest-southeast from the Sīrvān (Diyālā) River to Shīrāz.

Moreover in the early 18th century farmers had began to improve their livestock by selective breeding; notable among them was Robert Bakewell (1725-1795), a famous pioneer of selective breeding. During this period, there were other minor improvements such that the effects of this revolution are immediately apparent from the view point of yield data and the emergence of cities and civilizations.

1.4 Early agricultural sites in the world

Areas of domestication

Historic agricultural sites in the world arose as a result of the domestication of plant and animal, which distinguished the primitive hunter-gathering era. Although there are many scholarly debates about the details of these facts, it is widely recognized that there are five main areas/sites in the world in which domestication of plant and animals had arose:

1. Near East (Fertile Crescent)
2. Far East (South China around Yangtze River, & North China around Yellow River)
3. The new world (the Americas)
4. South-central Andes
5. Sub-Saharan Africa

The Near East (Fertile Crescent)

Some of the oldest sites of agriculture are in southwestern Asia, in the foothills around the area known as the Fertile Crescent, which today includes parts of Iran, Iraq, Turkey, Syria, Lebanon, and Israel. From sites such as Jarmo, Jericho, Ali Kosh, Cayonu, and others, remains of both plants and animals dated back 9,000 to 14,000 years.

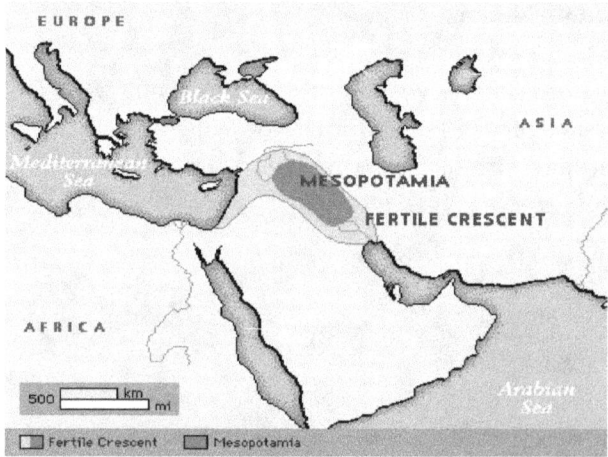

Figure 1-6: Map of the Fertile Crescent

There are recorded evidences of the hunter gatherers settling in small communities in the narrow band of land arcing across the Near East which was referred to as the Fertile Crescent.

The Far East

Archaeological evidences of excavations of dozens of sites in Asia indicated that agriculture may have first arisen at several locations in the Far East. These sites include the Yellow River and Yangtze River valleys in China (Figure 1-7).

Figure 1-7: The Far East

China was one of the first areas of the world where food production was invented. For instance, along the Yellow River in northern China, evidence two varieties of millet were being cultivated by 7500 B.P.[6], and dogs, pigs, and possibly cattle, goats, and sheep were domesticated by 7000 B.P. Also in the Yangtze River corridor of Southern China, rice was cultivated perhaps as early as 8400 B.P., and water buffalo, dogs, and pigs were domesticated by 7000 B.P.

The New World (the Americas)

Evidence of agriculture has also been found or rather invented independently in three regions of the New World 3, 000 to 4,000 years later than in the Middle East. These regions are the South American highlands (the south central Andes), the Eastern North America and the Tehuacan and Oaxaca valleys in Mexico (the Maya homeland called

---

[6] B.P. or (BP) represents 'Before Present'. Before Present' years are a time scale used mainly in geology and other scientific disciplines to specify when events in the past occurred. Meteorologists established 1950 as the origin year for the BP scale for use with radiocarbon dating, using a 1950-based reference sample of oxalic acid

Mesoamerica which spans five countries; Mexico, Guatemala, Belize, Honduras, and El Salvador).

Most of the initial archeological evidence for early agriculture in the New World has been obtained from the highlands of Mexico and Peru (Mesoamerica). Animal domestication was much less important in the New World than in the Old World. Animals domesticated in the New World include llamas (the largest domesticate), alpacas, guinea pigs, ducks, turkeys, and dogs. The large game animals hunted by early Americans either became extinct before they could be domesticated or were not domesticable.

Figure 1-8: Map of the Mesoamerican cities (LLC, 2000-2013)

In Mesoamerica, food production eventually led to early village farming communities. The earliest village farming communities developed around 3500 B.P. in the humid lowlands (the Gulf Coast of Mexico and the Pacific Coast of Mexico and Guatemala) and in the Mexican highlands (the Valley of Oaxaca). The gradual transformation of broad-spectrum foraging into intensive cultivation help laid foundation for the emergence of the states in Mesoamerica; about 3,000 years later than in the Middle East.

Between 10,000 and 4000 B.P, foragers in the Valley of Oaxaca had a broad-spectrum economy, exploiting a wide range of animals and plants. These foragers gathered a wild grass known as teocentli (or teosinte), the wild ancestor of maize. Between 7000 and 4000 B.P., harvesting and eventually cultivation brought about genetic changes in teocentli-maize (more kernels per cob, increased cob size, more cobs per stalk, tougher axes, softer husks).

Early farming in the Mexican Highlands (South-central Andes)

The Andes extend from north to south through seven South American countries of: Venezuela, Colombia, Ecuador, Peru, Bolivia, Chile, and Argentina. The ancient peoples of the Andes such as the Incas have practiced irrigation techniques for over 6,000 years.

Because of the mountain slopes, terracing has been a common practice. Terracing, however, was only extensively employed after Incan imperial expansions to fuel their expanding realm. The potato holds a very important role as an internally consumed staple crop. Maize was also an important crop for these people

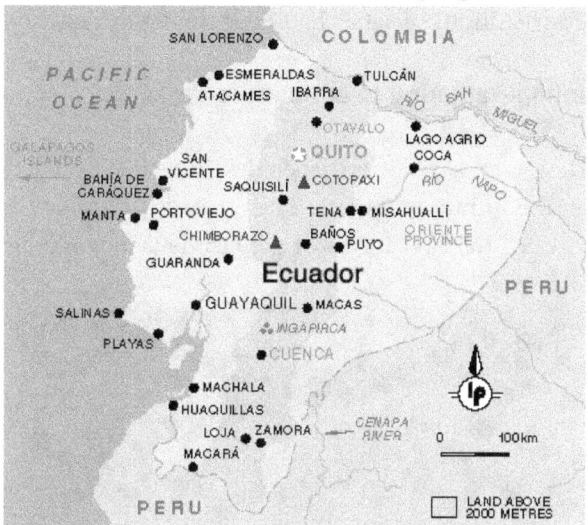

Figure 1-9: Map of the Andes cities

Agriculture in sub-Saharan Africa

Sub-Saharan Africa is, geographically, the area of the continent of Africa that lies south of the Sahara Desert excluding Sudan (regions below the separated boundary of Figure 1-10) . Agriculture in Sub-Saharan Africa refers to the agricultural system that is predominantly small-scale farming system with more than 50% of the agricultural activity performed by women, producing about 60-70% of the food in this region.

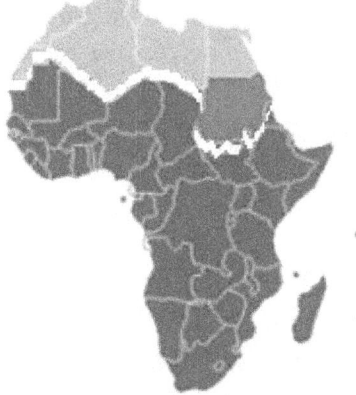

Figure 1-10: Map of the Sub-Saharan Africa (green areas)

Sub-Saharan Africa has more variety of grains than anywhere in the world. Between 13,000 and 11,000 BCE wild grains began to be collected as a source of food in the cataract region of the Nile, south of Egypt. The collecting of wild grains as source of food spread to Syria, parts of Turkey and Iran by the eleventh millennium BCE. By the tenth and ninth millennia, southwest Asians domesticated their wild grains, wheat and barley after the notion of collecting wild grains was spread from the Nile. In each of these regions, the number and sizes of agricultural societies had been on the increase.

Transition from hunting-gathering to agriculture

As the number and size of agricultural societies increased, they expanded into lands traditionally used by the hunter-gatherers. This process of agriculture-driven expansion led to the development of the first forms of government in some agricultural centers as mentioned above.

Figure 1-11: Hadza men return from a hunt

New evidences of the transition from Hunting-Gathering to Agriculture was largely hinged on population pressure which forced them to acquire more food on available space noting that early crops may have always been grown for food, and fiber and not for ceremonial purposes. Figure 1-11 showed two Hadza[7] men returning from a hunt.

Seed Culture in the Old and New Worlds

Seed culture was known to have originated in drier subtropics of both the northern and southern hemispheres around 8,000-10,000 years ago. Largely around the Tigris and Euphrates Rivers (Fertile Crescent), the Indus River (Northern India and Pakistan) as well as the Huang Ho (Yellow River-Yangtze) in the Northern China. Seed culture developed most rapidly in ecologically diverse regions especially in the earliest village farming communities in western Iran where wheat, barley, and domestic animals first appeared.

---

[7] Hadza men are one of the few contemporary African societies that live primarily by foraging. They are an ethnic group in north-central Tanzania, living around Lake Eyasi in the central Rift Valley and in the neighboring Serengeti Plateau.

There are evidences that equally suggested that earliest seed culture practiced in the Tehuacan Valley- Southern Mexico include corn, maize while the seed culture in the new world include maize-bean-squash complex among other crops.

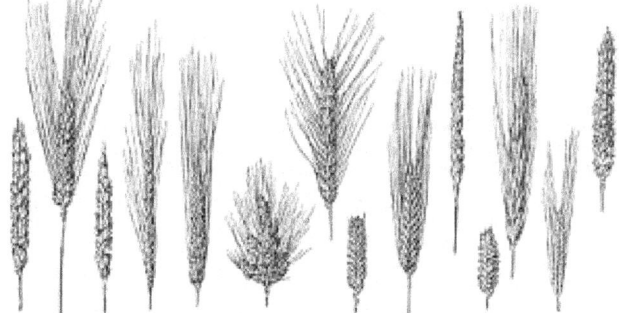

Figure 1-12: Ancestral wheat and barley

Three main staples (maize, potatoes, and manioc) were domesticated in the New World, along with other crops (e.g., beans, squash, quinoa, goosefoot, marsh elder, and sunflower) that added variety and nutrition to the diet. As seed cultures moved from the highlands to the valleys, irrigation practice developed, tillage systems developed and also selection of crop varieties improved

1.5 Plant and animal domestication

Essentially all of the arable land in the world is under cultivation. Yet agriculture began just a few thousand years ago, long after the appearance of anatomically modern humans. With the coming of large communities, families no longer cultivate the land for themselves and their immediate needs alone, but for strangers and for the future. They worked all day instead of a few hours a day, as hunter-gatherers had done (Pfeiffer, 1977).

Reasons for domestication

Attempt to explain why hunter/gatherers began to cultivate plants and raise animals gave rise to few satisfying answers and include:

1. Climatic change,
2. Population pressure,
3. Sedentism,
4. Resource concentration from desertification,
5. Girls' hormones,
6. Land ownership,
7. Geniuses, rituals,
8. Scheduling conflicts,

9. Random genetic kicks,
10. Natural selection,
11. Broad spectrum adaptation and

Domesticated plant & animal

In contrast to the Paleolithic or Old Stone Age (Old World, or pre-agricultural society), the *Neolithic* (New Stone Age, or agricultural society) cultures of the New World had domesticated an impressive array of plants but comparatively few animals.

Near East: Early plant domesticates in the Near East areas include einkorn wheat, emmer wheat, barley, pea, lentil, and vetch while dogs, goats, and sheep were among the domesticated animals.

Far East: Among the earliest plants domesticated in the Far East were rice, foxtail millet, broomcorn millet, rape, and hemp, with evidence of domesticated cattle, pigs, dogs, and poultry as well.

New World: Among the earliest crops domesticated in the New World were squash, corn, chili peppers, amaranth, avocado, gourds, beans, and both white and sweet potatoes, with only dogs, turkeys, llamas, alpacas, guinea pigs, and Muscovy ducks as domesticated animals.

Sub-Saharan Africa: Numerous crops have been domesticated in the region and spread to other parts of the world. These crops included sorghum, castor beans, coffee, cotton, okra, black-eyed peas, watermelon, gourd, and pearl millet. Other domesticated crops included enset, African rice, yams, kola nuts, oil palm, and raffia palm. Domesticated animals in the Sub-Saharan Africa include the guinea fowl and the donkey.

1.6 Agriculture: A revolution or an evolution?

Many theories have been proposed to answer the question of agricultural origination. Two views are widely propagated: evolutionary and revolutionary.

Revolutionary view

Some stated that agriculture was the discovery of a brilliant man who, with a flash of insight, realized

a. That if you sow seeds, the crop will grow, or
b. That plants growing in a medium, or dump sites, were growing from discarded seeds or

c. That seeds buried with the dead (as food for the afterlife) gave rise to plants at grave sites.

These theories viewed agriculture as a revolution; in fact, the term *Agricultural Revolution* was used to describe the transition from foraging to agriculture. It was suggested that this revolution spread quickly because agriculture was thought to be an improvement over the hunter-gatherer lifestyle in that a dependable food source could be easily grown rather than collected from the wild.

Evolutionary view

However, in the early 1960s, archaeologists questioned the revolutionary view, arguing that agriculture emerges as a result of gradual cultural evolution. They also reasoned that

a. Hunter-gatherers knew the wild plants, knew how they grew, and would incorporate farming along with foraging as part of an overall food-collection strategy when necessary. For example, certain aboriginal groups in coastal Peru abandoned their farming practices whenever fish became plentiful.
b. There was a transitional stage between simple foraging, in which small nomadic bands followed the wild plants and animals, and agricultural societies with their sedentary lifestyle. During this transitional stage, foraging groups formed settlements but sent out members to hunt and gather.

This more complex strategy resulted in changes in the social organization of the groups and permitted populations to increase. This transitional stage lasted for several thousand years in some locations until resource stress or environmental change led to the switch to agriculture.

Archeologists believed that in the Near East, for example, the climatic dry period around 11, 000 years ago brought about a change in the distribution of cereal grains (wheat and barley). Applying their botanical knowledge, these foragers gradually changed from collecting these wild cereals to cultivating them.

1.7 Agricultural evolution: History and perspective

The modern theory of the origin of agriculture is really an evolutionary one and not a revolutionary one. Most authorities agreed that agriculture arose independently in different areas over several thousand years, most likely as a natural consequence of intensified foraging. Records showed groups of hunter-gatherers had been seen in parts of south-west Asia, 23 000 years ago, to have started transformation of their settlement and subsistence strategies and develop large, permanently co-residential communities well before the beginning of agriculture (Trevor, 2010).

It was stated that the hunter-gatherers knew the wild plants, know how they grew, and began to incorporate farming along with foraging as part of an overall food collection strategy. Eventually, there was a transition between simple universal and largely unchanged for millennia (Lee & DeVore, 1968) foraging lifestyle where nomadic bands followed the plants around, to sedentary agricultural societies, where the people stayed in one place, and grew crops.

Figure 1-13: A hunter-gathering society

The foragers changed from collecting wild cereals to actually cultivating them. They began to gather, then cultivate and settle around, patches of cereal grasses and to domesticate animals for meat, labour, skins and other materials, and milk. This appears to have formed the basis of advanced civilization in both the old and new worlds.

The process of agricultural development in the world, both in the productive forces or in the adjustment of agricultural production relations, have demonstrated a gradual historical process of evolution, which embodies the historic evolutionary ages of agricultural development.

1.8 Agricultural revolution: History and perspective

The word revolution as used literarily indicates that there are sudden and radical changes that had occurred to agriculture over a period of time (Baker, 2009). However, changes that occurred in agriculture were radical but were far from sudden. So, obviously, use of the word revolution is applied with a certain license. In the so-called revolution, technology and climate played major roles. The technology in the form of an implement, an instrument or a process creating the revolution is usually relatively easy to cite. However, the part climate played is more difficult to pinpoint. Climate is like civilization itself, we can look back upon it and recognize changes that have occurred, but definitive explanations as to causes and timing remain elusive.

In his paper Evolutionary Fitness, Dr. Arthur De Vany (© 2005) wrote that: *The most important revolution in the history of Homo sapiens, was the agricultural revolution beginning about 10,000 years ago in Asia and near the Mediterranean.*

Prior to the domestication of plants, man was a hunter and a gatherer and most probably stored little food and measuring yield. A gatherer more or less unconsciously seeks plants that have larger seeds, more seeds per ear, and a compact inflorescence (Evans, 1980). The loose inflorescence of oats and the relative difficulty in gathering it may explain why oats was probably the last of the small grains to be domesticated.

The domestication of plants, possibly the greatest single milestone in man's history, is generally accepted as the point where plants retained their seed upon maturity. Thus, threshing is required as well as the intentional sowing of the grain, since the succeeding crop no longer occurs as a result of the natural shedding of the seed due to movement of the plant, for example, by the wind or passing animals. Over time, yields have been expressed in three different ways and a fourth may come into use in the near future. They are shown in Table 1-1.

Table 1-1: Crop yield expressions and approximate period of use

Expression	Period
Yield/human effort*	Pre-domestication of plants
Yield/seed sown	3-8,000 B.C. to 1000-1500 A.D.
Yield/land area	1000-1500 A.D. to present
Yield/critical factor	Future use?

First agricultural revolution

The first agricultural revolution and the associated dawn of civilization apparently coincided with the warming of the earth centered around 5,000-6,000 years ago, following the end of the Pleistocene (the last "ice age"). The ice age ended about 8,000-10,000 years ago. The climate change that occurred following this ice age was a definite improvement and undoubtedly played an important part in this first revolution. With the first agricultural revolution, a whole new set of conditions came into being which are generally associated with the dawn of civilization, Table 1-2.

Plant domestication and crop storage: With plant domestication, it was no longer possible to rely on naturally scattered grain for regeneration of the crops. Sowing became essential along with the necessary self-discipline required to hold some grain aside as seed for the next year. Along with this self-discipline came a greater awareness of the weather and its powerful influence on crop yields as noted, for example, in the famous Biblical account recorded in Genesis 41 of Pharaoh's dream, Joseph and his plans for the 7 full years and the 7 years of famine (according to Biblical scholars occurring between 2000-1700 BC).

Table1-2: Events of the first agricultural revolution

Features	Events
Climate	Post-glacial warm period
Wild plants	Domesticated plants
Hunters and gatherers	Gatherers and hunters
Nomadic life	Sedentary life
Wild animals	Domesticated animals
Weapons	Tools
Location	Tigris and Euphrates rivers
Occurrence	5 - 8,000 BC
Yield expression	Yield/human effort
Search for better lands (?)	Yes, as indicated by settlement locations

Yield measurement: A new criterion of determining yield was established during this era. That is, yield of grain compared to the amount sown. This is the measure mentioned in the Bible (parable of the sower and the soil, Matt 13:1-23, Mk 4:1-20 and Lk 8:4-15[8]), also confirmed by Romans and by ancient Chinese writers. At the time of Christ, wheat yields were about 3 or 4 to 1, but on good, fertile soils they could be considerably higher (Evans, 1980).

Development of agricultural implements: From the rudimentary beginnings, agricultural implements initially may have been used as weapons which served a secondary purpose as a tool to scratch the surface. Then, the hoe was well developed and were initially human powered, first there was the foot, also called "digging sticks" which was used as shown in Egyptian tomb paintings, and then the hand-ards (the Egyptian *mr*), which are pointed and strong enough to clear rocky soil and make seed drills was introduced.

Figure 1-14: Animal-drawn true ard ploughing

---

[8] Reference material: The Holy Bible KJV

There was a deliberation whether these instruments should be termed ploughs, since they lack two essential parts of such as the coulter and the moldboard, often used to differentiate it from similar instruments. However, the domestication of oxen in Mesopotamia and by its contemporary Indus valley civilization, perhaps as early as the 6th millennium B C., provided mankind with the draft power necessary to develop the larger, animal-drawn true ard (or scratch), thereby ushering in the second agricultural revolution.

The second agricultural revolution

The next method of yield measurement introduced us to the second great agricultural revolution which took place beginning about 500 or 600 A.D. (Burke, 1978) and is centered upon the Medieval periods. This second revolution provides us with a fascinating story that few people even in agriculture are aware of. Like the first revolution, it rests upon a favorable climate and upon two special features - an improved agricultural implement and the horse. Together, these three features created a revolution in agriculture that can be compared to the post-World War II agricultural revolution, though it extended over a much longer period.

Improved agricultural implements: Sometime in the sixth century A.D., a different agricultural implement appeared which carried two extremely important features: a knife called a "coulter," which could cut through heavy roots, and a mold-board which lifted the cut soil to one side. These two features serve to define a mouldplough and separate it from other instruments. These features, the knife and the mouldboard, wrought major changes in Medieval agriculture, especially in combination with the horse and the very essential horse collar which probably entered the scene a century or two later.

Horse development: With the introduction of the horse collar, horseshoe, and nails, a remarkable series of events occurred immediately it was discovered that the horse could be truly exploited, Table 1-3. The horse, in contrast to the ox, is 50% faster and has greater endurance, working two to three more hours per day (Gimpel, 1976).

Table 1-3: Events of the second agricultural revolution (medieval period)

Climatic warm period (1700-1200 A.D.)	
Stick	
Moldboard (6th century)	
Wheeled	
Horse collar (9th century)	
Horse	1. 50% faster
	2. Greater endurance
	3. Horseshoe + nails
	4. Hitch in line

	Yield -- yield per unit area		
Social changes	1. Cooperatives formed		
	2. Villages formed		
	3. Population increase		
Physical changes	1. Field drainage		
	2. Field preparation		
	3. Field shape		
	4. Farming system practice: 3-field system	a. Diet improvement	
		b. Spread of risk	
		c. More land	
	5. Soil texture factor	Finer soils can be easily worked	
	6. Deforestation activities		
	7. Crops mostly planted	Oats a universal crop	

The fact that technology and those who contribute to it are seldom recognized is a phenomenon of long standing in the western world (Gimpel, 1976). This was noted by Plato, and even Leonardo da Vinci felt the scorn of the intellectuals of his day who considered him little better than a manual worker or technician. The foundations of our present technologically oriented society were not laid in the Italian Renaissance or the English industrial revolution but in the medieval period. In many respects the medieval period has outshone even the Renaissance, which the conventional historian referred to as the real flowering age of man's intellect.

Based on the generally accepted view of medieval man it is difficult to realize that he was surrounded by machines (Gimpel, 1976). Water power (mills) was developed as it had not been by the Greeks and Romans. Medieval man also made use of wind power. Windmills may have been introduced from the plateaus of ancient Persia (modern Iran and Afghanistan) (Gimpel, 1976). The development of power sources demonstrates the mechanical ability of medieval man.

New ways to express yields were established to become yield per unit land area, bushels or pounds per acre, and so on. Thus, in a sense, the limiting factor became the amount of land that could be farmed.

The Medieval period was described until recently as the Dark Ages- the whole period between the end of classical civilization, (that is, Greek and Roman period) and the revival of learning in the 15th century. It is little known because of a peculiarity of many historians that is only belatedly being corrected. Modern historians now have to revise history, and the term Dark Ages has been dropped in favor of the medieval period, subdivided into Early, Middle, and Late.

The third agricultural revolution

From today's vantage point it is hard to believe that between the 8th century (the medieval period) and the middle of 18th century, there was little change in agriculture. The tools of farming basically are the same with little advancement in technology. For instance, the farmers of George Washington's day had no better tools than had the farmers of Julius Caesar's day; in fact, early Roman s were superior to those in general use in America eighteen centuries later. Sure, tractors were taking over from the horse, and the binder, reaper, and threshing machine were reducing the work required. The U.S. horse population did not reach its peak until 1914, and crop yields were not all that much better than they were in the medieval period of about 700 years earlier.

Tillage methods had not changed significantly: as the "minute man's" of 1775 and those of the 20th century scenes; an "ard" is still used in Italy, the scratch in Ecuador and Korea as later as 1951.

The quest for better tools, improved production process and increased productivity, lead to the British agricultural revolution in England. It began with a man named Jethro Tull (1674-1741). The third agricultural revolution was therefore a period of agricultural development between the 18th century and the end of the 19th century, which saw a massive and rapid increase in agricultural productivity and vast improvements in farm technology.

In the 17th century seed was sown by hand. The farmer simply scattered seed on the ground. However in 1701 Tull invented the seed drill. This machine dropped seeds at a controllable rate in the straight lines. A harrow at the back of the machine covered the seeds to prevent birds eating them. Tull also invented a horse drawn hoe which killed weeds between rows of seeds.

Figure 1-15: Jethro Tull's seed drill, 1701

Furthermore new forms of crop rotation were introduced. Under the old system land was divided into 3 fields and each year one was left fallow. This was, obviously, wasteful, as one third of the land was not used each year. In the 17th century the Dutch began to use new forms of crop rotation with clover and root crops such as turnips and swedes instead of letting the land grow fallow. (Root crops restored fertility to the soil). In the 18th century these new methods became common in England. A man named Charles 'Turnip' Townshend (1674-1738) did much to popularize the growing of turnips.

Turnips had another advantage. They provided winter feed for cattle. Previously most cattle were slaughtered at the beginning of winter because there was not enough food to keep them through the season. Now fresh milk and butter became available all year round.

Moreover in the early 18th century farmers began to improve their livestock by selective breeding. One of the most famous pioneers of selective breeding was Robert Bakewell (1725-1795). There were other minor improvements. On light soil farmers used marl (clay with lime content). Other farmers drained their fields with stone lined trenches. Manure has always been used as fertilizer but in the mid-18th century farmers began to build underground tanks to protect manure from the weather.

The third revolution was the most abrupt of the three. The delay in the application of the accumulated technology was caused by World War II and then the Korean Conflict. But the effects of this revolution are immediately apparent when viewing yield data. The first portion, 1866-1938 shows a yield averaging only about 30 bushels per acre, which is not much better than a very good yield in the medieval period (if corn had been grown then). This was followed by a 42 bushel per acre average yield from 1939-1951 when some of the new technology such as commercial fertilizers and hybrid corn began to be applied (Baker, 2011). This was then followed by a yield trend, 1952-present, that has shown an increase equaling nearly 2 bushels per acre per year as technology became fully adopted.

Fourth method of yield measurement

The third revolution may have run its course or it may have received a boost from biotechnology. But with or without the application of a new technology, a fourth method of yield measurement may be used in the near future. It is the ratio of yield to a critical factor other than land. As the critical factor in the past has gone from human effort, to the amount of seed sown, to the amount of land used, it may soon change, for example, to the nitrogen, the phosphorus, or the energy expended. Perhaps the best one would be an economic one, since it also requires a superior bookkeeping system. Thus, the next yield expression might become yield per amount spent-could this usher in the 4th agricultural revolution?

1.9 Phases of agricultural development

Agriculture is characterized by different stages of development. Some of the level of agricultural productivity from the perspective of the development is divided into ancient (primitive) agriculture, traditional agriculture and modern agriculture. Theorists of these stages of development of agricultural division of the world from different perspectives reflecting the characteristics of agricultural development phase. Based on this understanding, the agricultural development can be divided into primitive agriculture, traditional agriculture and modern agriculture.

Primitive (forager/hunters) agriculture

Primitive agriculture was practiced from the Neolithic Age to the Iron before the advent of agriculture (agricultural embryonic period). Primitive agriculture with slash and burn as the basic mode of production was applied to a certain extent, using wood, stone and other simple tools, fire and water and other means of production. However, human beings have begun to adapt naturally to the active intervention in the natural, from the existing access to natural food to purposefully produce human needs of food, especially the beginning of the domestication of wild animals and plants to achieve the transformation from gathering (foraging from the wild) to farming (cultivates a crop and harvests it), and hunting to animal husbandry changes. This marked the beginning of traditional agriculture.

Traditional agriculture

Traditional agriculture is characterized with the use of basic hand tools inherited from the iron to the pre-industrial agriculture age, as experienced 2000 years ago for self-sufficiency in agriculture. During this period, the utilization of humans, animals and metal fabrication technique employed in semi-mechanized production of tools, and the development of techniques for artificial application of organic manure to improve soil fertility, gained recognition. This period witnessed progress in the inventions to improve the crops and livestock trait technology, the creation of inter-cropping, inter-planting and other crop rotation cropping system, more and more workers from the Natural Science and the research results to get the appropriate skills, the ability to use and transform nature. Traditional agriculture involves the use of traditional method of farming.

Features of traditional agriculture

1. Technical stagnation,
2. Production growth limited to the traditional experience of workers
3. Characterized by a small amount of capital,
4. Low-level of technology,
5. Characterized by subsistence farming practice

Subsistence (farming) agriculture

The term subsistence agriculture refers to a self contained and self sufficient unit where most of the agricultural production is consumed and some may be sold in local market is sold. Subsistence agriculture is that type of agriculture in which crops grown are consumed by farmer and his family. Traditionally, low level of technology and household labour are used to produce on a small output.

Characteristics of subsistence agriculture

The main characteristics of traditional or subsistence agriculture in brief are as follows:

1. *Land use*: Traditional farms are very small usually only 1 to 3 hectares. The goods produced on these small farming units is used mainly for consumption of the family. The consumption survival considerations dominate the commercial ones.
2. *Labour*: Labour used per hectare tends to be high in traditional farming. Mostly the family labour works on the subsistence farms. However the traditional farms may hire some labour during the busy time of the year. Family farm members may and do supplement their income by working off the farm part during slack times.
3. *Power and transport*: In many countries including Pakistan livestock is the main source of power on the field for transport of products and carrying out processing tasks like grinding sugar cane. The level of technology mostly used is simple and less productive.
4. *Productivity and efficiency*: The subsistence farming or traditional farms are characterized by low of inputs which are mostly provided by the farmer himself. For example seeds, cow dung manure etc are not purchased by the farmers. Yields per hectare, production per person and overall productivity tend to be low.
5. *Rationality and risk*: The traditional farmers are economically rational. They can be motivated to raise standard of living. The subsistence farmers are not adverse to changes but proposed changes must fit in into their current farming operations. The traditional farmers are now mentally ready to take risks of using modern inputs into their small scale farming operations.

Transformation of traditional agriculture to modern agriculture is necessary to:

1. Achieve value orientation from subsistence to market-based agriculture
2. Changes in industrial structure from the partition type to linkage-type system
3. Shift in mode of operation from extensive to intensive,
4. Improve labour skills from production to management system.

To achieve this change, the government wants to improve infrastructure, promote institutional innovation, and strengthen support for conservation, building service system, focusing on the role of guiding and demonstrations.

Modern agriculture

Modern agriculture is a new stage in the history of agricultural development targeted at feeding the world's growing population with little or no cost to the Environment. Modern agriculture is capable of producing greater yields than ever before, but intensification of agriculture does come at a price. Modern agriculture involves the use of modern farm implements technologies as a means of production or operation of the overall agricultural management. Modern agricultural era marked the beginning of the industrial revolution in agriculture; the industrial revolution in agriculture is the gradual movement towards commercialization and market-oriented agriculture. During this period, agriculture is in a market economy framework and the extensive use of the results of modern industry, science and technology, capital and other modern factors of production increased significantly.

How modern agriculture developed

Before the 1800's, most persons were peasant farmers. At the turn of the 100 years, the agriculture work force was down to 36%. Similar tendencies were happening in other countries as well. People were beginning to proceed away from ranches to the towns to find occupations in factories. By 1900, new technologies and agriculture procedures throughout the 1800's decreased the requirement for ranch employees while increasing production. There were now more large ranches, and less little farms. Larger ranches directed the way in seeking new procedures in cultivating, fertilizing, livestock breeding, harvesting, and the use of new farming inventions.

Basic practices of modern agricultural systems

Modern agricultural systems have been developed with two related goals in mind:

1. To obtain the highest yields possible and
2. To get the highest economic profit possible.

In pursuit of these goals, six basic practices have come to form the backbone of production:

1. Intensive tillage,
2. Monoculture,
3. Application of inorganic fertilizer,

4. Irrigation,
5. Chemical pest control, and
6. Genetic manipulation of crop plants.

Each practice is used for its individual contribution to productivity, but when they are all combined in a farming system each depends on the others and reinforces the need for using the others. The work of agronomists, specialists in agricultural production, has been keyed to the development of these practices.

Intensive tillage practice: The soil is cultivated deeply, completely, and regularly in most modern agricultural systems, and a vast array of tractors and farm implements have been developed to facilitate this practice. The soil is loosened, water drains better, roots grow faster, and seeds can be planted more easily. Cultivation is also used to control weeds and work dead plant matter into the soil.

Monoculture: When one crop is grown alone in a field, it is called a monoculture. Monoculture makes it easier to cultivate, sow seed, control weeds, and harvest, as well as expand the size of the farm operation and improve aspects of profitability and cost. At the same time, monocultures tend to promote the use of the other five basic practices of modern agriculture.

Use of synthetic fertilizers: Very dramatic yield increases occur with the application of synthetic chemical fertilizers. Relatively easy to manufacture or mine, to transport, and to apply, fertilizer use has increased from five to ten times what it was at the end of World War II (1939-45). Applied in either liquid or granular form, fertilizer can supply crops with readily available and uniform amounts of several essential plant nutrients.

Irrigation technologies: By supplying water to crops during times of dry weather or in places of the world where natural rainfall is not sufficient for growing most crops, irrigation has greatly boosted the food supply. Drawing water from underground wells, building reservoirs and distribution canals, and diverting rivers have improved yields and increased the area of available farm land. Special sprinklers, pumps, and drip systems have greatly improved the efficiency of water application as well.

Chemical pest control: In the large monoculture fields of much of modern agriculture, pests include such organisms as insects that eat plants, weeds that interfere with crop growth, and diseases that slow plant and animal development or even cause death. When used properly, synthetic chemicals have provided an effective, relatively easy way to provide such control. Chemical sprays can quickly respond to pest outbreaks.

Genetic manipulation: Farmers have been choosing among crop plants and animals for specific characteristics for thousands of years. But modern agriculture has taken advantage of several more recent crop breeding techniques. The development of hybrid seed, where two or more strains of a crop are combined to produce a more productive offspring, has been one of the most significant strategies. Genetic engineering has begun to develop molecular techniques that selectively introduce genetic information from one organism to another, often times from much unrelated organisms, with a goal of capitalizing on specific useful traits.

The basic characteristics of modern agriculture

Modern agriculture has gradually moved towards commercialization- market-oriented agriculture. Its main features include:

1. High technology development and application,
2. Capital intensive, characterized by the use of high-stage capital, technology, agriculture.
3. *Market maturity*: during this period, the main purpose of production is not self-sufficient, but rather to provide goods for the market in order to achieve profit maximization.: the degree of market matures during the transformation,
4. *Wider use of advanced technology*. Advanced technology is a key element of modern agricultural development. For instance the mid-19th century agriculture and chemical techniques of developing organic fertilizers, hybrid corn, wheat and rice-based 'green revolution' have gradually penetrated into the agricultural germplasm resources, plant and animal breeding, crop cultivation, animal husbandry, soil nutrients, plant protection and other fields.
5. *Better industrial system*. Improved industrial system is an important symbol of modern agriculture. And modern means of production, production technology suited to agricultural development breaks through the traditional methods to the universal adoption of mutual integration of different areas, urban and rural economic and social development, agriculture, industry chain to greatly extend markets for agricultural products.
6. *The ecological environment*: Modern agriculture with the use of chemicals and energy (mainly oil) and a large number of consumption as the start of its development though has made tremendous achievements, but also brings resources to destruction, pollution and other outstanding problems.

Problems of modern agriculture

There are usually problems associated with modern agriculture benefits notwithstanding. Some basic problems include

1. Excessive tillage led to soil degradation,
2. The loss of organic matter,

3. Soil erosion by water and wind,
4. Soil compaction.
5. devastating pest outbreaks in large monoculture farms
6. Agricultural water users compete with urban and industrial use and wildlife.
7. Hybrid seed has contributed greatly to the loss of genetic diversity and increased risk of massive crop failure, as well as an increased dependence on synthetic and non-renewable inputs needed for maintaining high yield.
8. Genetically engineered crops have the same negative potential, especially as the selection process takes place less and less in the hands of farmers working in their own fields, but rather in far away laboratories.

Impact of modern agriculture on environment

During the latter half of the twentieth century, what is known today as modern agriculture was very successful in meeting a growing demand for food by the world's population. Yields of primary crops such as rice and wheat increased dramatically, the price of food declined, the rate of increase in crop yields generally kept pace with population growth, and the number of people who consistently go hungry was slightly reduced. This boost in food production has been due mainly to scientific advances and new technologies, including the development of new crop varieties, the use of pesticides and fertilizers, and the construction of large irrigation systems.

Impacts of modern agriculture on environment could be any of the following but not necessarily limited to

1. *Increased water or wind erosion*: Raindrops bombarding bare soil result in the oldest and still most serious problem of agriculture. The long history of soil erosion and its impact on civilization is one of devastation. Eroded fields record our failure as land stewards. Lowdermilk (1953) described several civilizations that collapsed because of erosion. At the end of his historical sweep of failed civilizations, warning that unless we want to experience a similar fate, we had better heed these lessons of the past and safeguard our soils.
2. *Depleted groundwater supplies in irrigated areas*: Adequate rainfall is never guaranteed for the dryland farmer in arid and semiarid regions, and thus irrigation is essential for reliable pro auction. Irrigation ensures sufficient water when needed and also allows farmers to expand their acreage of suitable cropland (Poster, 1985). Unfortunately, current irrigation practices severely damage the cropland and the aquatic systems from which the water is withdrawn and partially returns.
3. *Nutrient loading of water bodies,*
4. *Loss of genetic diversity*: As modern agriculture converts an ever-increasing portion of the earth's land surface to monoculture, the genetic and ecological diversity of the

planet erodes. Both the conversion of diverse natural ecosystems to new agricultural lands and the narrowing of the genetic diversity of crops contribute to this erosion.

5. *Pesticide contamination or livestock odours*: In nearly all respects agriculture became an industry, sharing with the traditional manufacturing industries the problems of waste byproducts disposal since the introduction of cheap inorganic nitrogen fertilizers. Beside the benefit of high production, there is the potential environmental pollution that these massive chemical inputs can cause.

Impacts of modern agriculture upon soil quality

Soil quality plays a fundamental role in ensuring viable and sustainable agricultural production.

The impacts on soil quality associated with agriculture include

1. *A decline in soil organic matter*: Soil organic matter forms the carbon store which is the fundamental element for living fauna in soils that also controls other chemical and physical processes.
2. *Soil compaction due to vehicular movement*: intensive agriculture with heavy machinery and large livestock units has also caused soil compaction which is related to factors such as soil texture, packing density, moisture content and plasticity. Compacted soils are also more prone to erosion due to greater ease of runoff associated with the lack of surface roughness and a reduction in tendency for water to seep through the soil profile.
3. *Erosion problems* initiated by mechanical manipulation of soil as well as vehicular movements.
4. *Soil biodiversity disturbance*: soil biodiversity acts as the engine that makes soil alive and functional in the many processes that it governs. Soil biodiversity is influenced by a wide range of factors including the physico-chemical environment (e.g. Ph, temperature, water); the supply and availability of soil plant-derived inputs; managed inputs and indigenous organic matter; and the soil habitat.
5. *Soil contamination from organic fuel*: the sources of soil contaminants can be from diffuse or point source pollution and can be broadly classified as nutrients, heavy metals or organic pollutants. These soil contaminants can influence soil biodiversity.
6. *Soil sealing and soil Stalinization*: soil sealing and Stalinization only affect a small component of agricultural land.

Shift from modern agriculture to sustainable agriculture

Agriculture has been a way of life and continuous to be the single most important livelihood of the masses. The performance of agricultural sector influences the growth of economy. With the use of modern scientific technology in agriculture, we become self sufficient in food production. As modern crop production technology has considerably raised output but also created lot of problems to us. Our natural resources are degraded and diminished. Quality of environment is adversely affected. Problem of land degradation, pesticide residue in food products, gene erosion, atmospheric and water pollution is increasing at alarming rate. Modern technology of agriculture has high exploitative in nature enhancing pollution and causing enormous damage to environment.

In order to take advantage of new technologies and practices, farming systems will need to be viewed as **ecosystems**, or agricultural ecosystems. By monitoring both the positive and negative impacts of modern farming practices, ecologically based alternatives can be developed that protect the health of the soil, air, and water on farms and nearby areas, lower the economic costs of production, and promote *viable* farming communities around the world. Organic agriculture, conservation tillage, integrated pest management (IPM), and the use of appropriate genetic techniques that enhance local adaptation and variety performance are a few of the possible ways of ensuring the sustainability of future generations of farmers

1.10 Agricultural revolution: History and perspectives

Every continent in the world has experienced a period of agricultural retardation throughout history. These retardations effectively limited the population which local territories could sustain over long periods of time. These retardations occur when food is unavailable for a period of time, usually exceeding a year, and the resources necessary to sustain large static populations - agricultural workers, money, and transportation - are not available. Since food shortages are usually "local" affairs, the ability to buy and transport food over much longer distances helped minimize the impacts of local food shortages led to the process of agricultural development or revolution.

Benefits of agricultural revolution

1. **Unprecedented population growth** Improvements and new crops in turn supported unprecedented population growth, freeing up a significant percentage of the workforce, and thereby helped drive the Industrial Revolution.
2. **Machinery acquisition**: Improvements in tools and conditions help farmers to purchase greater masses of metal tools, new machines, and other goods, with the same quantity of earnings.
3. **Improved living standards**: Living standards partially improved, for instance, as the much cheaper textile manufacturing brought about by the Industrial Revolution made available cheap washable cotton underwear that raised personal hygiene levels and slowed down the spread of gastrointestinal diseases such as cholera.

4. **Fertilizer industrial growth**: Increased understanding of the important chemicals need for proper plant nutrition allowed new fertilizers to be imported and made. Eventually the establishment of a chemical fertilizer industry made it possible to restore the chemicals lost in growing crops to maintain the needed enhanced food plant growth.
5. **Increased productivity**: New agricultural equipment like cultivators, s, threshers, mowers, combines and balers were invented and powered by oxen, horses, steam power, then gasoline or diesel as the farm machinery became mechanized and allowed the farmers to become ever more productive.
6. **Improved farm transportation system**: The farm transportation system moved from pack animals to carts and wagons pulled by oxen, to mules or horses to trucks (after about 1915).
7. **Continuity and improvement:** New technologies in power source supply, metal working skills, monetary systems, investors and capital became available for agricultural continuation and improvement.
8. **Labour saving**: Encourages increasing innovations in agricultural machinery and labour saving devices development.

British agricultural revolution

The British agricultural revolution describes a period of agricultural development in Britain between the 15th century and the end of the 19th century, which saw an increase in productivity and net output that broke the historical food scarcity cycles. This revolution occurred over a period of several centuries (more likened to an evolution than a revolution) and was preceded or closely duplicated by many countries in Europe and their colonies. One of the keys to the British Agricultural Revolution was the development of ways of keeping and improving the arable land in Great Britain to counteract the loss of the soil's plant nutrients in cropping a given area. Higher yielding land was added to higher yielding crops with more yield/acre. Farm workers using more productive tools and machinery produced more crops with fewer workers.

There were large variations in the location and time different agricultural innovations were introduced in the many different agricultural zones in Britain and the rest of the world. The Agricultural revolution picked up speed as the Industrial Revolution and the advances in chemistry produced the scientific knowledge, wealth and technology for a more systematic development of commercial fertilizers and new and more productive agricultural machinery. New crops like potatoes (introduced about 1600) and maize were introduced from the Americas improving the yield/acre of arable land.

Advances that helped the agriculture revolution

Agricultural revolution, industrial revolution and scientific revolution developed simultaneously. Without increasing amounts of food to feed the increasing city populations, and without the capital, tools, metals and increased agricultural markets, the industrial and scientific revolutions could not have proceeded.

Scientific and technical knowledge generated by the Industrial and scientific advances in science; engineering and elementary botany had equally encouraged the progression of the agricultural revolution in the world.

Each of the so called "revolutions" supported and advanced the other revolutions—they were intricately linked together. By the 19th century, technological and general scientific conditions were central in the flourishing of the following identifiable important advances in the agricultural revolution/practices:

1. Evolution of more professional farm management skills,
2. New improved crops of higher yield,
3. New crop rotation systems such as four-field crop rotation involving turnips and clover (plus others) made it less necessary to have so much land lie fallow had more yield per acre
4. Selective breeding of livestock for larger size and other desirable characteristics.
5. New improved livestock that could be fed though the winters gave larger yields of meat. Larger drives of livestock increased the available meat supply in the ever increasing cities.
6. New irrigation systems called water meadows allowed longer growing seasons on pasture land and increased yield from hay fields allowing more animals to be raised.
7. More capital investment,
8. Better agricultural education,
9. Mechanization of farm work from oxen, horses, steam power and then gasoline or diesel power,
10. Improved fertilizations: Improved use and new fertilizers in addition to "manure" helped maintain and improve the arable land or reclaim formerly "waste" land.
11. Transportation systems flourished, too. Sailing ships, paddle steamers (after 1830), steam ships (after 1860), refrigerator ships (after 1890), toll roads (after 1700), canals (after 1760), railways (after 1830) allowed people, goods, foods and animal crops to be gathered and shipped cheaply and with increasing rapidity over ever increasing distances.
12. New food preservation techniques like canning (after 1800), refrigeration (after 1880) as well as traditional food storage techniques like root cellars and granaries were improved. These allowed more food to be stored over longer periods of time and minimized the impact of local shortages.

13. New technologies like the telegraph (after 1830) and the telephone (after 1880) greatly improved the speed and flow of information needed to keep up with prices, shortages and surpluses to know which crops to buy, ship or grow.

Effects of agricultural revolution on history

Sound advice on farming began to appear in England in the mid-17th century, from writers such as Samuel Hartlib, Walter Blith and others (Thirsk, 2008), but the overall agricultural productivity of Britain started to grow significantly only in the period of the Agricultural Revolution. It is estimated that the productivity of wheat was about 19 bushels per acre in 1720 and that it has grown to 21-22 bushels in the middle of the eighteenth century. It declined slightly in the decades of 1780 and 1790 but it began to grow again by the end of the century and reached a peak in the 1840s around 30 bushels per acre, stabilizing thereafter (Snell, 2014 modified).

The Agricultural Revolution in Britain proved to be a major turning point in history. The population in 1750 reached the level of 5.7 million. This had happened before: in around 1350 and again in 1650. Each time, either the appropriate agricultural infrastructure to support a population this high was not present or plague or war occurred (which may have been related), a Malthusian catastrophe occurred, and the population fell. However, by 1750, when the population reached this level again, an onset in agricultural technology and new methods without outside disruption, and also the effects of sugar imports, allowed the population growth to be sustained.

The increase in population led to more demand from the people for goods such as clothing. A new class of landless labourers, products of enclosure, provided the basis for cottage industry, a stepping stone to the *industrial revolution*. To supply continually growing demand, shrewd businessmen began to pioneer new technology to meet demand from the people. This led to the first industrial factories. People who once were farmers moved to large cities to get jobs in the factories. The British Agricultural Revolution not only made the population increase possible, but also increased the yield per agricultural worker, meaning that a larger percentage of the population could no longer work in agriculture but could and/or had to work in these new, post–Agricultural Revolution jobs.

Towards the end of the 19th century, the substantial gains in British agricultural productivity were rapidly offset by competition from cheaper imports, made possible by advances in transportation, refrigeration, and many other technologies into the 20th century.

1.11 The green revolution

The term "Green Revolution" was first used in 1968 by former United States Agency for International Development (USAID) director William Gaud, who noted the spread of the new technologies:

"These and other developments in the field of agriculture contain the makings of a new revolution. It is not a violent *Red Revolution* like that of the Soviets, nor is it a *White Revolution* like that of the Shah of Iran. I call it the *Green Revolution* (Gaud, 1968)."

Green Revolution refers to a series of research, development, and technology transfer initiatives, occurring between the 1940s and the late 1960s, which increased agriculture production worldwide, particularly in the developing world, beginning most markedly in the late 1960s (Hazell, 2009). The initiatives, led by Norman Borlaug, the "Father of the Green Revolution" credited with saving over a billion people from starvation, involved the development of high-yielding varieties of cereal grains, expansion of irrigation infrastructure, modernization of management techniques, distribution of hybridized seeds, synthetic fertilizers, and pesticides to farmers.

Figure 1-16: Norman Borlaug, father of the green revolution

For instance, India began its own Green Revolution program of plant breeding, irrigation development, and financing of agrochemicals in 1961 when at the brink of famine. India soon adopted IR8 – a semi-dwarf rice variety developed by the International Rice Research Institute (IRRI) that could produce more grains of rice per plant when grown with certain fertilizers and irrigation and in 1968, findings showed that IR8 rice yielded about 5 tons per hectare with no fertilizer, and almost 10 tons per hectare under optimal conditions. This was 10 times the yield of traditional rice (De Datta, et al., 1968).

This trend soon spread across the world through the supports of the World Bank; co-sponsored by the FAO, IFAD and UNDP which on 19th May 1971, founded the

1.12 History of agricultural development in Nigeria

This article was a comprehensive and in-depth account of historical background of agricultural development in Nigeria culled from an article authored by Nwachukwu Chinweizu .N. (aka Flash) titled: Discuss in details the history of agriculture in Nigeria from the colonial era to the present day, pointing out clearly all agricultural programmes. Posted to the web: 5/12/2006

The history of agricultural development in Nigeria is intertwined with its political history. This is discussed broadly in the context of the varying constitutional frame works, viz: Colonial, the internal self government and the post-1960 periods.

Crop production

The period of the colonial administration in Nigeria, 1861-1960, was punctuated by rather ad hoc attention to agricultural development. During the era, considerable emphasis was placed on research and extension services. The first notable activity of the era was the establishment of a botanical research station in Lagos by Sir Claude Mcdonald in 1893. This was followed by the acquisition of 10.4 kms of land in 1899 by the British Cotton Growing Association (BCGA) for experimental work on cotton and named the experimental area Moor Plantation in Ibadan.

In 1912, a Department of Agriculture was established in each of the then Southern and Northern Nigeria, but the activities of the Department were virtually suspended between 1913 and 1921 as a result of the First World War and its aftermath. From the early 1920s to the mid 1930s, there was a resurgence of activities and this period has been called the 'Faulkner Strip Layout' era in honour of the Director of Agriculture, Mr. O. T. Faulkner, who devised a statistical design for experimental trials in green manuring, fertilizer projects, rotational cropping systems and livestock feeding. From the late 1930s to the mid-1940s, there were significant intensification and expansion of research activities, and extension and training programmes of the Agricultural Departments. Additional facilities for training of junior staff in agriculture were provided, as well as scholarships for agricultural students in Yaba Higher College and Imperial College of Tropical Agriculture in Trinidad.

The intensification of hostilities during the Second World War (1939-45) led to the slowing down of activities and the call to Departments of Agriculture to play increasing roles in the production of food for the army and civilians in the country and the Empire. Production of export crops like palm products and rubber which could not be obtained from Malaysia as a result of Japanese war activities in South-East Asia, and such food items as sugar, wheat, milk, eggs, vegetables, Irish potatoes and rice whose importation was prevented by naval blockade of the high seas increased. A special production section of the Department of Agriculture was set up to deal with the situation.

Agricultural research activities

On the research side, attention was devoted largely to the possibilities of evolving permanent systems of agriculture that were capable of replacing rotational bush-fallowing systems prevalent in the country and realizing the promises of mixed farming in the north. During this period, the WAIFOR (West African Institute for Oil Palm Research) in Benin was started and the research on cocoa was intensified at Moor Plantation, Owena near Ondo and at Onigambari near Ibadan.

Achievements of the period include the development of 'Alien Cotton' in the south; rice cultivation in the Sokoto, Niger, Ilorin, Abeokuta Colony and Ondo provinces; the introduction of wheat cultivation in the more northern parts of the northern provinces; the expansion of production of such export crops as cocoa, oil palm and groundnut; development of agricultural implements as well as designing farm buildings; intensification of horticultural activities; the development of a marketing section of the Department; the extension of the Produce Inspection Service to cover all principal export crops; investigations into the possibilities for organized land settlement schemes; and investigations into the possibilities of irrigation in northern Nigeria.

The period of Internal Self Government, 1951-1960 began with the regionalization of the Departments of Agriculture in 1951, with a Director and an Inspector-General of Agriculture in each region. By October 1954, the post of Inspector General of Agriculture was abolished as a result of constitutional developments which led to independence of the Regional Departments. The Federal Department of Agricultural Research was retained since constitutional provisions placed agricultural research on the concurrent legislative list, while extension work remained a regional responsibility.

The research findings of the Federal Research Stations were to be transmitted through Regional ministries responsible for agriculture and natural resources. There was also the setting up, in 1955, of a Technical Committee of the Council of Natural Resources made up of Federal and Regional Ministers and officials for the formulation of nation al research programmes as well as the co-ordination of Federal and Regional research activities.

Regionalization of agriculture created a great awareness of the need for intensification of activities in both the research and extension fields. This led the Regions to expand, considerably, their research and extension activities in agriculture. The post-1960 was one of extensive planning and regional competition in agriculture. Concentration of attention on commodity exports, the utilization of taxation policy by the Marketing Boards as an instrument of development finance, and the belief that food production activities could take care of themselves without any governmental intervention, became the official farm policy. Under regional independence, the agricultural history of the nation entered a new phase of modification of traditional practices, in view of the incapacity of food production to meet the needs of the rising population and the inability of producers to reinvest in land.

These maladies were worsened by the inability of the then Federal Government to play a leading role in the nation's agricultural modernization. Before the middle of the 1960s, a Federal Ministry of Agriculture and Natural Resources was set up, and a phase of consolidation and co-ordination of projects for agricultural development began. In 1966, Federal initiative and control of the nation's agriculture were set in motion. This step in the right direction became more manifest with the creation, in 1967, of 12 States and the increased efforts to evolve a coordinated perspective for agricultural development in Nigeria.

Livestock development activities

Livestock production in Nigeria was dominated by nomadic pastoralism long before the advent of the British Colonial Administration. The immediate interest of the colonial government in livestock was with the health and hygiene of the domesticated cattle. Thus, the Nigerian Veterinary Department was established in 1914 with its head quarters at Zaria. In 1924, a small veterinary laboratory was established in Vom for the production of rinderpest serum.

Increased field services raised the demands on the laboratory hence the production of vaccines and other biological products was added to the functions of the laboratory. The recognition of the advantages of Vom as the centre for veterinary research and for vaccine production, coupled with the major emphasis on the health aspects of live stock production, led to the transfer of the head quarters of the Nigerian Veterinary Department from Zaria to Vom.

In October 1927, proposals for the establishment of a Stock Farm were made to the Government. The stated objective was 'to turn out, by purely selective breeding, male stock for use as stud by native stock owners.' It was proposed that three breeds, namely, the White Fulani, Gudali and Shuwa represented by a dairy herd of about 20 heads each be stocked at Shika. By 1934, it was fairly certain that either sweet potatoes or cassava could be fed to the cattle, as sources of energy.

At about that time too, it was realized that there is a heavy and growing export of cattle of the hoof from the North, and the intensity of this demand naturally fluctuates with the price of southern produce. Thus with price ruling high for palm oil in the South and low for cotton in the North, a prospective farmer is subjected to variations in his costs and returns. Such variations urged the planners to introduce the 'Mixed Farming Policy.' The policy was typified by the importation of six pullets and one cockerel of the Rhode Island Red breed from England in 1933 to Agege where crops like maize, cassava, yam, oil palm, kola, coffee, pineapple and citrus fruits were already cultivated.

The role of educational advancement in agricultural development in Nigeria was given prominence at an earlier stage. The value of an elementary education in the three Regions to farmers was appreciated and it was suggested that the introduction of a new interest into farming, such as the production of livestock in the Southern Provinces of Nigeria, would attract more educated youths into agriculture. A scheme was started in Katsina Province for teaching sons of farmers the best husbandry methods. Instructions were essentially practical in nature and were centered on mixed farming.

Similarly, the study of management of livestock was introduced to the lbadan Agricultural School where the Education and Agricultural Departments cooperated to train both teachers responsible for the management of school farms and the agricultural assistants for the Department of Agriculture. By 1938, three Conferences of West African Agricultural Officers had been held. Besides, the numerous attempts made between 1924 and 1938 to introduce fodder and browse plants into Nigeria (especially at the Veterinary Station, Vom, and the Agricultural Station at Samaru and at the Stock Centre at Shika) were reviewed.

The need for concerted effort at pasture and grassland management and improvement was adequately documented and a call for more cooperation between the livestock farmers and the traditional agriculturists was made. This was the beginning of organized efforts towards range management for livestock improvement in Nigeria. In 1940, milk-buying units were established in areas of the Jos Plateau and butter was produced on commercial scale. The production of cheese and bacon was undertaken shortly after and this became intensified during the Second World War.

After the War, livestock produce assumed consider able importance, while in 1948 the operations were taken over by the Department of Commerce and Industry. A Veterinary School was established at Vom in the early 1940s to train Nigerians for animal health work. A Livestock Investigation Centre (LIC) was also set up as auxiliary to the school and laboratory. Later, an Egg Production Unit was created to supply fertile eggs for virus research, vaccine for both the Veterinary and the Medical Departments and Poultry for research work and vaccine testing.

The Nigerian Veterinary Department played a very prominent role in the early history of livestock development in Nigeria. Indeed, by the end of the 1939-45 War, the Department had become internationally recognized and requests were made by the administration of most of the other West African Territories to the veterinary laboratory in Vom for the supply of vaccines. The serious nature of trypanosomiasis (sleeping sickness) in man and animals was also of great concern to the Colonial Administration in the West African Territories and the need to control this disease led to the establishment in 1947 of a West African Institute for Trypanosomiasis Research (WAITR).

A main laboratory to study the animal was sited in Vom-on the Jos Plateau, an ideal location since the tsetse fly vector was absent in that area. Prior to 1951, the Nigerian Veterinary Department had its headquarters, laboratories and a school in Vom, with field offices in each Region. With the coming of regional governments, the Nigerian Veterinary Department was split into separate regional departments. The Director of Veterinary Services became the Inspector-General of Animal Health Services, while the designation of the regional heads remained the same, except for that of Northern Region which was changed to Director of Veterinary Services.

The post of Inspector-General carried executive authority in the regions only in so far as matters connected with hides and skins trade was concerned. In October 1954, with the introduction of a new Constitution, the Regional Departments became completely autonomous. The post of the Inspector-General of Animal Health Services was re-designated as the Director of Veterinary Research, responsible to the Federal Government and with executive authority on veterinary matters in Lagos.

In 1967, when 12 States were created in Nigeria, each state assumed responsibility for veterinary matters, within its boundaries. The initial breeding policy designed to improve livestock in Nigeria concentrated on the locally available breeds of animal. About 1950, there was a modification of this policy, whereby exotic breeds of cattle were introduced to upgrade the local stock. The Western Nigeria Development Corporation (WNDC) established the Upper Ogun Ranch for the commercial production and distribution of cattle. In the Eastern Region, South Devon cattle were introduced at the Obudu Ranch. Friesian bulls were imported to the farm at Agege in Lagos; the Teaching and Research Farm at the University of Ibadan obtained foundation stock of cattle from Shika. Extensive facilities were also established for research in piggery and poultry.

The administrative machinery for agricultural development and co-ordination was also modified. Technical committees established for the various aspects of primary production were modified. The Veterinary Technical Committee was replaced by the enlarged National Livestock Development Committee which reported to the National Council for Agriculture and Natural Resources. The Livestock Meat Authority, established to serve the northern states, had recently been empowered to act on a national scale in collating data and con ducting surveys as well as in researching into various aspects of livestock production, slaughter and marketing in Nigeria.

Fisheries development activities

The history of fisheries development in Nigeria is a comparatively recent one, although reports have shown that a fishing company operated from the coastal waters of Lagos long before 1915. Deliberate efforts at developing the country's fisheries can be said to date back to the Second World War when, because of the naval blockade of the high seas, the then Colonial Administration decided to develop the country's local resources, including fisheries.

A fisheries organization was established in 1941 as a Fisheries Development Branch of the Agricultural Department of the Colonial Office and a Senior Agricultural Officer was appointed to conduct a survey of the industry and its possibilities. The headquarters was sited at Apese village and later at Onikan in Lagos, from where, assisted by a part-time voluntary officer, preliminary experiments in fish culture in brackish water ponds at Onikan were carried out and surveys were conducted on the canoe fisheries of Apese village and Kuramo waters around Victoria Island, Lagos.

A small fisheries school was also established at Onikan. Early in 1945, the Fisheries Development branch was temporarily transferred from the Agricultural Department to the Development Branch of the Secretariat. A Fisheries Development Officer was appointed and a Five-Year Plan for Fisheries Development was formulated and incorporated in the Ten-Year Plan of Development and Welfare in Nigeria, laid on the table of the Legislative Council on 13th December, 1945. From this date to 1947, the Branch became a section of the Department of Commerce and Industries with a Principal Fisheries Officer in charge.

In addition to the brackish water fish culture experiments and canoe fisheries surveys, other activities were initiated. Small motor fishing crafts were acquired for exploratory fishing in the estuaries, lagoons and creeks. It was considered 'that these fisheries should receive priority treatment at this stage in Nigeria over sea fisheries'. This was in spite of the earlier reports on the fishing company which showed that suitable trawling grounds existed off Lagos at depths of 18-65m.

Other activities undertaken included tests of rice growing in tidal mangrove swamps, where such an activity could be combined with fish farming, and improvements in the social conditions of the wholly fishermen populations of two small villages in Lagos. Between 1948 and 1950, major efforts were made at extending the artisanal fisheries programme to other coastal areas of Nigeria. An active extension service was established to demonstrate the benefits of improved fishing techniques and gear to the coastal canoe fishermen.

In addition, trawling surveys were undertaken in the vicinity of Lagos and Cameroon's and a sub-station was maintained at Opobo for several years before it was closed down due to lack of funds and personnel. A start was also made in fish culture in inland areas by the construction of experimental ponds and the stocking of the then existing ponds and reservoirs. A Fish Farmer was appointed to extend this aspect of production and this culminated in the establishment in 1951, of a 160ha industrial-scale fish farm at Panyam on the Jos Plateau. By the end of this period, the branch had grown to become the Federal Fisheries Services under the Federal Ministry of Economic Development.

Between 1952 and 1957, the bulk of the marine biological research was performed by the West African Fisheries Research Institute (WAFRI) at Freetown, Sierra Leone; a unit was maintained at Birnin Kebbi to conduct research into the fisheries of River Sokoto. In consequence of Nigeria's and Ghana's withdrawal of their support, the WAFRI was disbanded with effect from 31st March 1957; the fisheries research activities of the Federal Fisheries Service were expanded to take care of this function. Under the 1954 Constitution of Nigeria, the fisheries organization was split between the Federal and Regional Governments.

The Federal Fisheries Service of the Federal Ministry of Economic Department was headed by a Director with laboratories and headquarters in Lagos. The Western Region Fisheries Division of the Ministry of Agriculture and Natural Resources was headed by a Principal Fisheries Officer. Its headquarters and offices were at Ibadan and a Sea Fisheries Section at Lagos, a Marketing and Distribution Section at Warri, Organization and Inspectorate at Epe and Fish Culture Section at Ibadan and Asaba. The Eastern Region Fisheries Division of the Ministry of Agriculture was under the charge of a Principal Fisheries Officer and the headquarters at Aba and an outstation at Opobo.

The Fisheries Section of the Ministry of Agriculture of the Northern Region was under the charge of a Senior Fisheries Officer while the headquarters was located first at Baga and later at Malarnfatori, Lake Chad. In addition, the Northern Region Fish Farm at Panyam was placed under the administration of the Region's Ministry of Trade and Industry, and was under the charge of a resident Fish Farmer. The Federal Fisheries Service had the constitutional responsibility for fisheries development and research in the Lagos Federal Territory and research in any other part of the country where the Regional Government invited it to carry out any specific research activity.

In practice, however, the Western Region Sea Fisheries Section in Lagos, sited in the same compound as the Federal Fisheries Service, catered for all fishermen whatever their origin and whether they actually lived in Western Region Territory or in the Federal Territory. So the Federal Fisheries Service left all Lagos fisheries development work to the Western Region fisheries Division. It concerned itself, instead, with the development of the modern fishing vessels (trawlers) including their licensing; the planning of a fishing terminal for Lagos; and also with research. The Regions never requested Federal assistance towards research. They either tried to conduct research themselves or asked for international multilateral (FAO/UNDP) or bilateral (USAID) help.

Thus, the Federal Fisheries Service had to, on its own initiative, identify regional research needs and carry out what studies it felt were needed. On such initiatives, the Malarnfatori station was established on the Lake Chad; the brackish water fish-farming project was developed at Buguma; and studies were initiated at the Kainji Dam site. The period 1956-66 witnessed great expansion in Nigeria's fishing activities. In the coastal trawler fleet, from a single registered trawler in 1956, the fleet was built up, by 1960, to 13 while the total fish catch increased ten-fold during the period.

This level of production was sustained up to 1963 but catches fell in 1964-66, following heavier exploitation of the Lagos fishing grounds. By this period, however, commercial quantities of prawns had been discovered in the eastern parts of the country and many of the vessels converted to prawn fishing, thus reducing the pressure on the fish stock. By1970, the fishstock had fully recovered and the expansion of inshore fishing activities was becoming so rapid that plans were then made to regulate fishing in order to conserve the rather limited resources. The period also saw a considerable increase in the artisanal fisheries. This has been attributed to the concentration of fishing activities close to the rich grounds; higher money returns for efforts; general improvement in processing, storage and distribution methods; improvement in the type of fishing craft used and, especially, to the higher gear efficiency due to a complete changeover to synthetic fiber.

The general result was that the contribution of fisheries to the country's QDP quadru pled between 1960 and 1970. Fisheries Service left all Lagos fisheries development work to the Western Region fisheries Division. It concerned itself, instead, with the development of the modern fishing vessels (trawlers) including their licensing; the planning of a fishing terminal for Lagos; and also with research. The Regions never requested Federal assistance towards research. They either tried to conduct research themselves or asked for international multilateral (FAO/UNDP) or bilateral (USAID) help. Thus, the Federal Fisheries Service had to, on its own initiative, identify regional research needs and carry out what studies it felt were needed. On such initiatives, the Malarnfatori station was established on the Lake Chad; the brackish water fish-farming project was developed at Buguma; and studies were initiated at the Kainji Dam site.

The period 1956-66 witnessed great expansion in Nigeria's fishing activities. In the coastal trawler fleet, from a single registered trawler in 1956, the fleet was built up, by 1960, to 13 while the total fish catch increased ten-fold during the period. This level of production was sustained up to 1963 but catches fell in 1964-66, following heavier exploitation of the Lagos fishing grounds. By this period, however, commercial quantities of prawns had been discovered in the eastern parts of the country and many of the vessels converted to prawn fishing, thus reducing the pressure on the fish stock.

By1970, the fish stock had fully recovered and the expansion of inshore fishing activities was becoming so rapid that plans were then made to regulate fishing in order to conserve the rather limited resources. The period also saw a considerable increase in the artisanal fisheries. This has been attributed to the concentration of fishing activities close to the rich grounds; higher money returns for efforts; general improvement in processing, storage and distribution methods; improvement in the type of fishing craft used and, especially, to the higher gear efficiency due to a complete changeover to synthetic fiber. The general result was that the contribution of fisheries to the country's QDP quadrupled between 1960 and 1970.

Agricultural development since Independence

The 1962-1968 development plan was Nigeria's first national plan. Among several objectives, it emphasized the introduction of more modern agricultural methods through farm settlements, co-operative (nucleus) plantations, supply of improved farm implements (e.g. hydraulic hand presses for oil palm processing) and a greatly expanded agricultural extension service. Some of the specialized development schemes initiated or implemented during this period included:

1. Farm Settlement Schemes; and
2. National Accelerated Food Production Programme (NAFPP), launched in 1972.

There were also a number of agricultural development intervention experiments, notably

1. Operation Feed the Nation, launched in 1976;
2. River Basin and Rural Development Authorities, established in 1976;
3. Green Revolution Programme, inaugurated in 1980; and
4. The World Bank-funded Agricultural Development Projects.

While each of the above programmes sought to improve food production, the ADPs represented the first major practical demonstration of the integrated approach to agricultural development in Nigeria. The experiment which started with World Bank funding, with projects at Funtua (1974), Gusau (1974) and Gombe (1974), blossomed into Ayangba (1977), Lafia (1977), Bida (1979), Ilorin (1980), Ekiti-Akoko (1981) and Oyo-North (1982) agricultural development projects. Following successful negotiations for multi-state agricultural development projects with the World Bank, each state of the country, and the federal capital, Abuja, now has one ADP.

The years since the early 1960s have also witnessed the establishment of several agricultural research institutes and their extension research liaison services. Some of the major institutions are:

1. Agricultural Extension and Research stock production and fisheries production in Nigeria Liaison Service (AERLS) at the Ahmadu in recent years, is presented in the next two chap Bello University, Zaria, established in 1963;
2. The International Institute of Tropical Agriculture (IITA), at Ibadan and;
3. International Livestock Centre for Africa (ILCA)

CHAPTER 2

AGRICULTURAL SOCIETY
The historic developments

2.0   Society in the middle ages

The rulership

In the Middle Ages society was like a pyramid. At the top of the pyramid was the king. Below him were the barons; while at the bottom of society were the peasants. The king ruled by divine right. In other words people believed that God had chosen him to be king and rebellion against him was a sin. However this system proved awkward for that did not stop rebellions! A great deal depended on the personality of the king. If he was a strong character he could control the barons. If he were weak or indecisive the barons would often rebel.

The lifestyle

Most people in the Middle Ages lived in small villages of 20 or 30 families in the countryside and made a living from farming. The land was divided into 3 huge fields; each year, two of the three fields were sown with crops while one was left fallow (unused) to allow it to recover. Most peasants owned only one ox so they had to join with other families to obtain the team of oxen needed to pull at the implement. Later, the land was sown by men and women planted peas and beans.

The livelihood

Most peasants also owned few cows, goats and sheep. Cows and goats gave milk and cheese. Most peasants also kept chickens for eggs. They also kept pigs. Peasants were allowed to graze their livestock on common land. In the autumn they let their pigs roam in the woods to eat acorns and beechnuts. However they did not have enough food to keep many animals through the winter. Most of the livestock was slaughtered in autumn and the meat was seasoned with salt to preserve it.

The settlement

The peasant's homes were simple wooden huts with wooden frames filled in with wattle and daub (strips of wood woven together and covered in a 'plaster' of animal hair and clay). However in some parts huts were made of stone either whitewashed or painted in bright colours. At night in summer and all day in winter the peasants shared their huts with their animals. Parts of it were screened off for the livestock. Their body heat helped to keep the hut warm.

The adaptability

In the early 14th century the climate of the world cooled and there were a series of famines, the population began to fall. So many people died and there was a serious shortage of labour. Parliament tried to fix wages by law to prevent them rising but this was impossible to enforce. By the 15th century the system of serfdom or villeinage had broken down in England.

2.1 Agricultural Society

An agricultural society refers to any group or organization where the chief occupation devoted to the improvement of agriculture. The aim of agricultural society is to promote scientific development in agriculture and its membership composition includes farmers, researchers, scientists, engineers etc.

Over the centuries, agricultural societies spread into those environments that could be easily adapted to agriculture, and foragers gradually became restricted to marginal areas. By the late twentieth century, foraging societies had largely disappeared, constituting only a tiny percentage of the human population and limited to a few tropical rain forests, deserts, savannas, tundra, and boreal forests

2.2 Characteristics of agricultural societies

1. Cultivation of land through the as this invention enabled the people to make a great leap forward in food production.
2. It increased the productivity of land through the use of animals and bringing to the surface the nutrients of the soil. Combining irrigation techniques with the use of the increased the productivity and the crop yield.
3. It also brought fallow land under cultivation.

4. The size of the agricultural societies increased as it lessened the burden of large number of people who engaged themselves in other activities.
5. Agricultural societies lead to the establishment of more elaborate political institutions like formalized government bureaucracy assisted by the legal system.
6. It also leads to the evolution of distinct social classes -those who own the land and those who work on the other's land. Land is the major source of wealth and is individually owned. This creates major difference between the social strata.
7. Agricultural societies provide the basis for the establishment of economic institutions. Trade becomes more elaborate and money is medium of exchange.
8. It also demands the maintenance of records of transaction, crop harvest, taxation, governmental rules and regulations.
9. Religion becomes separate institution with elaborate rituals and traditions.
10. The agricultural societies support the emergence of arts and cultural artifacts due to surplus food production people tend to divert their attention to other recreational activities.
11. There is far more complex social structure; population size increases, cities appear, new institutions emerge, social classes arise, political and economic inequality becomes inbuilt into the social structure and culture becomes much more diversified and heterogeneous.

2.3 Features of early agricultural societies

Early agricultural societies were characterized by the hunter-gatherers society, the agrarian society and the agricultural revolution society. Each of these societies was characterized by certain features describing the mode of livelihood, association, activities and relationships. Each of these societies is discussed in the following sections.

2.3.1 Hunter-gatherer society

A hunting and gathering society comprises of a group of men and women with distinct responsibilities as they continue to live within their enclaves. Women in the hunting and gathering society were like today's stay-at-home-moms except that they gather small things like grains, berries, and other fruits around their dwellings per time. They took care of the kids. The fathers and men alike hunted animals like sheep, cow, and wolfs. Every now and then the fathers took the boys out to teach them how to hunt so that they can be strong when they grow up.

Characteristics of hunter-gatherer society

1. These societies also had extensive knowledge of the fauna (animals) and flora (plants) unique to particular areas.

2. They developed basic tools to help them hunt and gather, and to utilize their resources.

Figure 2-1: Hunter-gatherer society (the Neolithic man)

Technology in hunter-gatherer society

By 10,000 BC, humans had a range of technologies to aid them in their exploitation of the environment. The most fundamental of these was the ability to make and maintain fire. Fire played an important part in the mythologies of later societies - the Greeks told the story of Prometheus, the great benefactor of mankind, stealing fire from the gods. This suggests that humans invested this capability with great reverence, tinged with fear.

Fire was certainly of enormous significance to their lives. It gave them warmth and light, extending their geographical habitat to the colder latitudes as well as into dark environments such as caves. It enabled them to continue communal life after nightfall, and must therefore have strengthened their ability to tell stories round the hearth – a key element in human culture. Fire allowed people to cook their food, thus expanding their source of nutrition to less digestible or tasty plants. It was also used to harden wooden spears, making it possible to kill larger animals.

Figure 2-2: A hunter-gatherer society (Wind River Mountains of Wyoming, 1870)

The hunter-gatherer people of 10,000 BC used stone, wood, bone and antlers for their weapons and implements. Some groups practiced primitive mining, or more strictly quarrying, for flint, digging shallow pits and trenches. People wore clothing made from

animal skins, which they sewed together using intricately-crafted bone needles. They had mastered the use of cords and threads fashioned from plant materials to aid them in making their clothes as well as for making baskets.

Their weaponry included spears, bows and arrows, and harpoons. This last brought the food resources of lake, river and shore within their grasp, and indeed coastal peoples ventured some distance out to sea in small boats made from reeds or logs. They had already domesticated one species of animal, the dog (probably around 15, 000 BC), which they used for hunting.

Hunter-gatherer family groups

The ancient hunter-gatherers lived in small groups, normally of about ten or twelve adults plus children. They were regularly on the move, searching for nuts, berries and other plants (which usually provided most of their nutrition) and following the wild animals which the males hunted for meat.

Each group had a large "territory" over which it roamed – large, because only a small proportion of the plants in any given environment were suitable for people to eat, and these came into fruit at different times of the year meaning a large area of land was needed to meet the food needs of a small number of people. The group's territory had regular places where it stopped for a while. These might be caves or areas of high or level ground giving them a good all-round vision of approaching animals (and hostile neighbours), and where they would build a temporary encampment.

Hunter-gatherer clan

These family groups mentioned above belonged to larger "clans" of 50 to 100 adults, spread over a wide area and whose members regarded selves as a "people", descended from a common ancestor. Kinship was crucially important. This more than anything else gave them their identity and defined their place in the world. More practically, it told them who their friends and allies were, and governed whom they could or could not marry (incest, though differently defined at the margins, was a universal taboo, but marriage outside the clan was also restricted). Myths gave them their world view – how the universe was born, how humans came to be and so on – and there is clear evidence for spiritual beliefs, and indeed for belief in some kind of life after death.

There may well have been individuals within clans particularly revered for their wisdom and judgment, or even credited with special magical powers; but it is highly unlikely that anyone exercised any significant authority over any group larger than the family group. There were no kings or chiefs in such societies, and, because people ate what their own small group gathered or killed, no one was richer or poorer than his neighbours.

2.3.2 The agrarian society

The invention of tools marked the beginning of agrarian societies 6000 years back. According to Collins dictionary of Sociology, Agrarian society refers to any form of society especially those traditional societies primarily based on agricultural and craft production rather than industrial production. Agrarian societies have also been described as those employing animal drawn implements to cultivate land. The mode of production (i.e. cultivation) of the agrarian society distinguishes it from the hunter-gatherer society which produces none of its food.

Main features of agrarian society

1. The economy of this society is based on agriculture and the society is divided into a number of classes based on the feudal system.
2. Agrarian society viewed children as economic assets where large families were necessary for survival.
3. The system of production is based on the use of manual power.
4. Means of transport and communication are underdeveloped.
5. A system of joint family prevails, since a number of hands are required on the fields.

2.3.3 Agricultural revolution society

Some societies of 10,000 BC already had distinctive styles of art. These ranged from crude patterns on their weapons and tools, through modelled clay figurines of animals and women (presumably fertility spirits), to the wonderful sequence of cave paintings of animals and mysterious symbols found in south western France and northern Spain, dating from 3,500 BC to 9,000 BC.

Figure 2-3: Iron Age

The agricultural revolution society was characterized by domestication and cultivation of plant and animals as well as sophistication of tools and equipment for agricultural production.

Domestication

Domestication can be defined as the human modification of a plant/animal – one that is identifiably different from its wild ancestors and its extant wild relatives. In short, domestication involves genetic change through conscious or unconscious human selection. The move from shifting agriculture to domesticated agriculture was preceded and made possible by the millennia of accumulated experience of wild plants and animals, and trial-and-error experimentation.

Characteristics of domesticated plants

The stages of harvesting, planting and storing imposed various artificial selection pressures such as the following:

1. Plants with favoured characteristics are preferentially harvested
2. Plants preferentially harvested are resown

In more detail, some of these selection pressures involved the following:

1. Plants provided with a seed bed of open soil encounter diminished competition
2. Harvesting and resowing of larger clusters of seed heads
3. Single harvesting event
4. Seeds with larger food reserves germinate quicker
5. Quick germination confers competitive advantage, and reduced need for protective seed coat and dormancy.

Over time, these selection pressures produced changes in the crop and seeds that are characteristic of domesticated crops. These changes (referred to as domestication markers) are most pronounced when comparisons are made between the domesticated crop and its wild relatives.

Figure 2-4: Animal and plant domestication

Characteristics of domesticated animals

Galton (1822 - 1911) identified behavioural and physiologic characteristics of animals which would make them better candidates for domestication i.e. pre-adaptations to domestication:

1. Hardy, flexible, generalist feeding habits; easily adjusting to new conditions of disease, temperature and confinement
2. A liking for humans
3. Comfort-loving
4. Useful
5. Breed freely - fewest and least constraining behavioural, situational cues for reproduction
6. Easy to tend social and roaming animals capable of group interactions
7. Gregarious, social groups of both sexes, maintain a dominance hierarchy, and are thus predisposed to submission. e.g. goats and sheep are placid, slow-moving foragers, not territorial and form highly social groups with a single dominant leader.

Cultivation

Cultivation involves the deliberate sowing or other management of plants which do not necessarily differ from wild populations. There had been a gradual shift from collecting to cultivation with continued reliance on hunting and gathering. Finally there was almost complete reliance on agriculture as the major source of nutrition.

Consequences of cultivation food production

Consequent on land cultivation and food production are the followings

1. Increased carrying capacity of the land
2. Development of sedentary societies
3. Changes in social structure
4. Craft specialization
5. Civilization

CHAPTER 3

ENGINEERING SOCIETY
The Historic developments

3.0 Introduction

The historic developments within engineering society are targeted at exploring the complex interactions between agriculture, technology and society human side of engineering. Agricultural society has been discussed in the previous chapter. This chapter explores the inter-relationships between engineering and the society as well its involvements in historic agricultural development.

3.1 Engineering society

Engineering society is an umbrella association formed by the coming together of several members of professional institutions with the primary objective of providing a forum where members of the profession can interact to share ideas. This body provides a wider platform for its operation than the professional institution. An engineering society is a professional organization for engineers of various disciplines. Some are umbrella type organizations which accept many different disciplines, while others are discipline-specific.

Many award professional designations, such as corporate member of Nigerian Society of Engineers (MNSE), European Engineer, Professional Engineer, Chartered Engineer, Incorporated Engineer or similar. There are also many student-run engineering societies, commonly at universities or technical colleges.

3.2 Status of engineers in society

The perennial question 'what is the status of engineers in society?' has been an object of debate since the mid 1800s. The question itself is problematic as the very definition of an engineer is still being hotly debated. It must be understood however that engineers vitally contribute to global challenges; and their hard work and innovation directly powers the success of the global markets.

In general the public do neither appreciate nor understand the vital contribution that engineers make towards the development of society. One might argue that this lack of knowledge may be a result of too few engineering role models around as well as the politics of professional development. Among a cross section of children aged 14 to 18 that wanted to pursue a career in engineering many were unable to name a famous engineer as role model.

Increasingly shocking is the unfortunate reality that the government does not recognize or consult the professional engineering institutions before making decisions. This furthermore reduces the status of engineers. Comparing this situation with many of the other professional disciplines such as Medical or Legal profession, their institutions are approached and consulted before government legislation is put in place.

3.3 History of agriculture and engineering

The history of interrelationship between agriculture and engineering is the biggest story of our time. It is a story of free entrepreneurship, perseverance and endurance of the individual, of vision, idealism and cooperation among men, of the lightening of human toil and the release of millions of workers from farms to feed the ever hungry industrial revolution. By no means least, it is the story of producing food necessary to win two global wars, keep our allies alive and millions of the defeated enemy from starvation (Worthington, 1966).

As early as 1915, steam traction engine had attained its highest development as the forerunner of the farm tractor rather than the predecessor. The *steam traction engine* was regarded as the instrument of expansion; while the tractor is regarded as the instrument of progress. The invention of the tractor, followed by only sixteen years Otto's practical embodiment application of the Beau de Rochas power cycle to a heat engine, marked the advent of a new order - the *age of power farming* (Worthington, 1966).

The pattern of progress follows that of the geometric progression; the scientist discovers the principles, the engineer applies the principles to agricultural development. The engineers are thereafter referred to as agricultural engineers and the profession, agricultural engineering.

3.4 Historic development of Agricultural Engineering Society

Before the emergence of agricultural engineering, engineering problems relating to agriculture were tackled by adapting the knowledge from fields of architecture, civil and mechanical engineering to solving such problems. The world outlook on agricultural engineering development historically started in the United States of America in the early 19th century. In 1906, a three-man forum attended by F. R. Crane of the university of Illinois, Jay Brownlee Davidson (referred to as the *"father of agricultural engineering,"*) of the then Iowa State College (ISC) now Iowa State University and C. A. Ocock of the university of Wisconsin, met at the university of Illinois to discuss teaching techniques and development of instructional materials for the discipline marked the beginning of engineering knowledge to solving problems on the farm.

Prior to the 1906 forum, in 1905, J. B. Davidson had developed the first professional agricultural curriculum at Iowa State College (ISC). Courses offered include; agricultural machines, agricultural power sources, with an emphasis on design and operation of steam tractors; farm building design; rural road construction; and field drainage. Davidson also becomes the first president of the American Society of Agricultural Engineers in 1907, leading agricultural mechanization missions to the Soviet Union and China.

J B Davidson

Also in 1906, three persons had shown keen interest in the profession and their activities actually prepared the groundwork for the 1906 forum. These individuals are; Dr. Elwood Mead, an agriculturist and a civil engineer who at one time was in charge of the US Bureau of Reclamation is one of the earliest advocates for agricultural engineering as a distinct profession, Profs. O. V. P. Stout and C. R. Richards who had been teachers of agricultural engineering and practical mechanics respectively at the school of agriculture of the University of Nebraska as far back as 1896.

Following the success of the 1906 forum, a second meeting was held in 1907 with an enlarged attendance at the agricultural engineering building of the University of Wisconsin in Madison. At the meeting, American Society of Agricultural Engineers (ASAE) was formally organized: a constitution and by-laws were adopted for the society and society was initiated as a National Engineering Society. Pioneer executive members of the society were then elected as follows:

1. Jay Brownlee Davidson   President
2. C. A. Ocock   First Vice President
3. F. R. Crane,   Second Vice President
4. L. W. Chase   Secretary
5. W. M. Nye   Treasurer

With this development, agricultural engineering as a distinct profession came into being. Departments of Agriculture Engineering were established in some universities across the USA. Iowa state university is reported to be the first university to award a first degree of

Bachelor of Science in Agricultural Engineering in 1910. By 1925, there were ten institutions and by 1950, the number of institutions awarding degrees in agricultural engineering had risen to 41 and so spread to other parts of the world.

Every country now has a Society of Agricultural Engineering. There are now European Society of Agricultural Engineers, Asian Association of Agricultural Engineers, and Euro-Asian Association of Agricultural Engineers among others. The umbrella organization of all these societies is the International Commission of Agricultural Engineering (CIGR), with headquarters in Belgium. In Africa, there is the South and Eastern Africa Society of Agricultural Engineering and also the West African Society of Agricultural Engineers

Agricultural engineering education in Nigeria

Agricultural engineering education in Nigeria started in the colleges of agriculture and polytechnics, with training in farm mechanization with emphasis on tractorization which was later expanded to cover all aspects of agricultural engineering. Agricultural engineering training was substantially done overseas until about the early 1960s when local opportunities started to be available. In the early days of agricultural engineering services, the first set of Nigerian agricultural engineers was produced through the re-training of professionals in civil engineering and agronomy.

Prior to the 1960s, the expertise and services of agriculturists and civil engineers were used to solve engineering problems on Nigerian farms. The interest and challenges of engineering services on the farm, made some of them to seek for opportunities to retrain themselves in what today forms the agricultural engineering curriculum. Some of the pioneer agricultural engineers in Nigeria then were therefore also specialists of other disciplines.

Gradually, the relevance of agricultural engineering began to be appreciated both by the government and those engaged in agricultural practices and local opportunities for full-fledged training of agricultural engineers were considered desirable. Local training of agricultural engineers in Nigeria started with the teaching of parts of the present day agricultural engineering curriculum to students in the schools and colleges of agriculture such as in Akure, Ibadan and Zaria, and faculties of agriculture in some universities. Some of such graduates were awarded degrees, diplomas and certificates in agricultural mechanization but not agricultural engineering (Mijinyawa, 2005).

Historic development of Nigeria Society [Institution] of Agricultural Engineers

Attempts to have a forum for agricultural engineers in Nigeria dated back to 1965. That year, Professor Cargill of the University of Nigeria, Nsukka, Mr. Hewitt of BEWAC Nigeria Limited, Dr. Layide Onafeko, Engr. Deji Osobu and Professor F. O. Aboaba came together to form the *Nigerian Society of Agricultural Engineers (NSAE)*. Mr. Hewitt was elected chairman while Professor Aboaba became the secretary. The forum then decided to include non-graduates such as technologists with National Diploma or certificate in agricultural engineering as members. Such members included Messrs Achike, Otuyalo and Solagbade.

The society took off with regular holding of meetings and held the first conference in Benin City in 1967. It was during the conference that the state of Biafra [9] was declared which marked the beginning of the Nigerian civil war from 1967 to 1970. The Nigerian civil war prevented the society from functioning until 1975 when with the incoming of a crop of young and dynamic Nigerian agricultural engineers resuscitated the society with Professor Ayo Makanjuola as president. The society effectively took off once again with the following objectives:

i. To promote the science and art of engineering in agriculture.
ii. To encourage agricultural research.
iii. To foster and promote agricultural engineering education.
iv. To advance in every possible ways the standards of agricultural engineering.
v. To promote the intercourse of agricultural engineers among its members and with allied technologies.
vi. To encourage the enhancement of professional competence of its members.

Membership of the society is opened to those who have either undertaken or are undergoing a professional Agricultural Engineering Curriculum and others who may not necessarily be Agricultural Engineers by training but who in the course of their employment or research have contributed or are contributing to the advancement of Agricultural Engineering. Between when it was founded and 1999, the institution existed as a society and was then known as the Nigerian Society of Agricultural Engineers (NSAE).

Considering the benefits derivable from being a division under the umbrella organization for the engineering profession in Nigeria; a merger with the Nigerian Society of Engineers (NSE) was, a formal merger agreement was signed on the 27th July 1999 between NSE and NSAE. Upon signing the agreement, the name was changed to The Nigerian Institution of Agricultural Engineers (NIAE) and it became a **Division of the NSE** with national secretariat within the premises of the National Centre for Agricultural Mechanization (NCAM), Ilorin, Kwara state. The Institution organizes annual conferences, which have transformed into an international event. Since 2003, the Institution also started an annual public lecture series. The Institution has a journal, the Journal of Agricultural Engineering Technology, which is published biannually and a quarterly newsletter.

---

[9] Sectionalization of the then Eastern Nigeria Region as a republic

CHAPTER 4

AGRICULTURAL ACTIVITIES
Historic developments

4.0   Introduction

Farm/or agricultural operations include, but are not limit to, activities such as the raising and harvesting of crops, raising of poultry or animals, logging, and forestry operations. In more concise term, "*Agricultural operations*" means

1. The growing or harvesting of crops from soil (including forest operations), and the raising of plants at wholesale nurseries, but not retail nurseries, or the raising of fowl or animals for the primary purpose of making a profit, providing a livelihood, or conducting agricultural research or instruction by an educational institution, or
2. Agricultural crop preparation services such as crop storage, crop processing, and poultry and animal care activities. These activities are limited to only the first processing activities after harvest, and not subsequent processing, canning, or other similar activities.
3. For forest operations, agricultural operation services include milling, peeling, producing particleboard and medium density fiberboard, and producing woody landscape materials.

4.1   Agriculture in the biblical era

The various agricultural operations regarded as common, practiced in the biblical times are described under the following operational activities as in the biblical era.

Land preparation

Ancient biblical records indicated that little or no preparation for sowing is done before swinging of seeds on the plains, but in the hilly regions, the larger stones, which the tilling of the previous season has loosened and which the winter's rains have washed bare, are picked out and piled into heaps on some ledge, or are thrown into the paths, which thus become elevated above the fields which they traverse.

Sowing/seed planting

If grains are to be planted, the seeds are scattered broadcast by the sower. If the land has not been used for some time, the ground is first ploughed, and when the seed has been scattered, it is ploughed again. The sower may keep his supply of seed in a pocket made by pulling up his outer garment through his girdle to a sufficient extent for it to sag down outside his girdle in the form of a loose pouch. He may, on the other hand, carry it in a jar or basket as the sowers are pictured as doing on the Egyptian monuments. As soon as the seed is scattered it is ploughed in before the ever-present crows and ravens can gather it up. The path of the plough in the fields of the hilly regions is a tortuous one because of the boulders jutting out here and there (Matt 13: 3-8) or because of the ledges which frequently lie hidden just beneath the surface (the rocky places of Christ's parable).

When the ploughman respects the footpaths which the sufferance of the owner has allowed to be trodden across his fields or which mark the boundaries between the lands of different owners, and leaves them unploughed, then the seed which has fallen on these portions becomes the food of the birds. Corners of the field where the plough cannot reach are hoed by hand.

Reaping

After the ploughing is over, the fields are deserted until after the winter rains, unless an unusually severe storm of rain and hail (Ex 9:25) has destroyed the young shoots. Then a second sowing is made. Toward the end of May or the first week in June, which marks the beginning of the dry season, reaping begins. Whole families move out from their village homes to spend the time in the fields until the harvest is over. Men and women join in the work of cutting the grain. A handful of grain is gathered together by means of a sickle held in the right hand (Deut 16:9). The stalks thus gathered in a bunch are then grasped by the left hand and at the same time a pull is given which cuts off some of the stalks a few inches above ground and pulls the rest up by the roots. These handfuls are laid behind the reapers and are gathered up by the helpers, usually the children, and made into piles for transporting to the threshing-floor.

Threshing

Threshing-floors are constructed in the fields, preferably in an exposed position in order to get the full benefit of the winds. If there is a danger of marauders they are clustered together close to the village. The floor is a level, circular area 25 to 40 ft. in diameter, prepared by first picking out the stones, and then wetting the ground, tamping or rolling it, and finally sweeping it. A border of stones usually surrounds the floor to keep in the grain.

The sheaves of grain which have been brought on the backs of men, donkeys, camels, or oxen, are heaped on this area, and the process of tramping out begins. In some localities several animals, commonly oxen or donkeys, are tied abreast and driven round and round the floor. In other places two oxen are yoked together to a drag on which the driver, and perhaps his family, sits or stand, are driven in a circular path over the grain, the bottom of which is studded with pieces of basaltic stone. There are biblical evidences of threshing instruments (Isaiah 28:27) on cart wheels powered by horses (Figure 4-1).

Figure 4-1: Threshing machine in operation

In other districts an instrument resembling a wheel harrow is used, the antiquity of which is confirmed by the Egyptian records. The supply of unthreshed grain is kept in the center of the floor. Some of this is pulled down from time to time into the path of the animals. All the while the partly threshed grain is being turned over with a fork. The stalks gradually become broken into short pieces and the husks about the grain are torn off.

Winnowing

This mixture of chaff and grain must now be winnowed. This is done by tossing it into the air so that the wind may blow away the chaff. When the chaff is gone then the grain is tossed in a wooden tray to separate from it the stones and lumps of soil which clung to the roots when the grain was reaped. The difference in weight between the stones and grain makes separation by this process possible. The grain is now poled in heaps and in many localities is also sealed. This process consists in pressing a large wooden seal against the pile. When the instrument is removed it leaves an impression which would be destroyed should any of the grain be taken away.

This allows the government offers to keep account of the tithes and enables the owner to detect any theft of grain. Until the wheat is transferred to bags someone sleeps by the pries on the threshing-floor (Ruth 3:2, 7). If the wheat is to be stored for home consumption it is often first washed with water and spread out on goats' hair mats to dry before it is stored in the wall compartments found in every house. Formerly the wheat was ground only as needed. This was then a household task which was accomplished with the hand-mill or mortar

Care of vineyards

No clearer picture to correspond with vine culture than that mentioned in (Isa 5: 1, 6). Grapes probably served an important part in the diet of Bible times as they do at present. In the season which begins in July and extends for at least three months, the humblest peasant as well as the richest landlord considers grapes as a necessary part of at least one meal each day. The grapes were not only eaten fresh but were made into wine in wine presses. No parallel however can be found in the Bible for the molasses which is made by boiling down the fresh grape juice. Some writers believe that this substance was meant in some passages translated by wine or honey, but it is doubtful. The care of the vineyards fitted well into the farmer's routine, as most of the attention required could be given when the other crops demanded no time.

Raising of flocks

The leaders of ancient Israel reckoned their flocks as a necessary part of their wealth (Genesis 13:2). When a man's flocks were his sole possession he often lived with them and led them in and out in search of pasturage (Psalm 23; Matt 18:12), but a man with other interests delegated this task to his sons (1 Samuel 16:11) or to hirelings. Human nature has not changed since the time when Christ made the distinction between the true shepherd and the hireling (John 10:12). Within a short time of the writing of these words the writer saw a hireling cursing and abusing the stray members of a flock which he was driving, not leading as do good shepherds.

The flock furnished both food and raiment. The milk of camels, sheep and goats was eaten fresh or made into curdled milk, butter or cheese. More rarely was the flesh of these animals eaten. The peasant's outer coat is still made of a thawed sheepskin or woven of goats' hair or wool.

4.2   Farming in the early agricultural era

Early agriculture had been generally characterized by evolution into two major types of farming; traditional farming method and mechanized farming methods, which represents the basis for classification of different levels of mechanization in agricultural operation.

The characteristics and operations of each method are described in the following subsequent sections.

4.2.1 Traditional farming activities in the early agricultural era

Traditional farming is characterized by

1. The use of fire for clearing a new farm plot (slash-and-burn agriculture),
2. Superficial tillage by hand,
3. Often planting on mounds or ridges,
4. Mixed cropping using a number of carefully composed crop associations,
5. The lack of any inputs for fertilizing and crop protection.

Slash and burn agriculture

Slash and burn is a method of agriculture primarily used by tribal communities for subsistence farming (farming to survive). It is a process of cutting down the vegetation in a particular plot of land, setting fire to the remaining foliage, and using the ashes to provide nutrients to the soil for use of planting food crops. Humans have practiced this method for about 12,000 years, ever since the transition known as the Neolithic Revolution, the time when humans stopped hunting and gathering and started to stay put and grow crops. Today, between 200 and 500 million people, or up to 7% of the world's population, uses slash and burn agriculture.

The cleared area following slash and burn, also known as swidden, is used for a relatively short period of time, and then left alone for a longer period of time so that vegetation can grow again. For this reason, this type of agriculture is also known as shifting cultivation.

Burning of the bush has the following advantages:

1. It kills most grasses and weeds, so that the first weeding needs to be done relatively late. The alkaline ash raises the pH and availability of cations in the surface soil.
2. Nutrients like P, K and Ca are made available in soluble form, which stimulates the growth of plants during the early growing, period.

On the other hand, clearing by fire has been heavily criticized for destroying the soil. Here, in less technical terms, are some disadvantages of burning:

1. The carbon, nitrogen and sulphur in the fallow and litter are lost in the burn (but not the amounts in the soil humus).
2. The fire damages trees like oil palms, raffia palms, etc.

3. There is a build-up of fire-tolerant, low-productive species.

Burning is, however, still an indispensable part of modern agricultural system. Crops in bush fallow rotations depend on the pH effect of the burn, and in grass fallow systems the burning of the grass is still considered to be more advantageous than composting - in particular in terms of return on labour (Lagemann, 1977).

Clearing in the forest zone

Burning of grass and undergrowth is very common in the forest zone, even more so the burning of the taller trees. In most cases, the dead trees are left standing on the farms, though sometimes the trees are allowed to continue to grow as they are used for shade. We may distinguish between the following systems of bush clearing in the forest zones:

1. Burning of the grass and trees
2. Use of grass and undergrowth as manure, burning off the trees
3. Use of grass and undergrowth as manure and felling of the trees for timber or firewood.

Clearing in the grassland area

In the grassland areas, we have identified four distinct methods of farm clearing:

1. Burning of grass and shrubs.
2. Burning according to the nkara[10]-system.
3. Covering the grass with soil for green manure.
4. Leaving the grass on top of the soil for manure.

Tilling

There are several traditional methods of tilling. The simplest one, practiced on fertile, well drained soil, is the simple scratching of the surface with a cutlass, planting maize seeds and covering them at once with soil. Tilling and planting are done at the same time. According to the slope of the land, the farmers often make mounds or ridges when tilling. These ridges, mounds or beds often contain grass and the remains of the previous crops as manure. Initially, ridges were made down the slope of a hill. In this way, work was easier for the women who worked moving uphill. Contour ridging means a much more

---

[10] Nkara-system is a traditional agricultural system of clearing land in the grassland area involving the cutting down of grass, covering it with soil to form a ridge, and then burning the grass. This keeps all the ash in the soil and makes a slow burning fire. It is on nkara ridges that egusi melons and Irish potatoes would be grown in the first year after clearing.

uncomfortable working position, the more so the steeper the slope. The downward sloping ridges make erosion worse.

Hoe-farming

The basic idea of soil scratching for weed control is ancient and was done with hoes or mattocks for millennia before cultivators were developed. In the overwhelming majority of cases tilling is done by hand. A hoe with a large blade is the universal tool for tilling. When agriculture was first developed, simple hand-held digging sticks and hoes were used in highly fertile areas, such as the banks of the Nile where the annual flood rejuvenates the soil, to create drills (furrows) to plant seeds in.

Digging sticks, hoes, and mattocks were not invented in any one place, and hoe-cultivation must have been common everywhere agriculture was practiced. Hoe-farming is the traditional tillage method in tropical or sub-tropical regions, which are characterized by stony soils, steep slope gradients, predominant root crops, and coarse grains grown at wide distances apart. While hoe-agriculture is best suited to these regions, it is used in some fashion everywhere. Instead of hoeing, some cultures use pigs to trample the soil and grub the earth.

Plough farming

The primary purpose of ploughing is to turn over the upper layer of the soil, bringing fresh nutrients to the surface, while burying weeds, the remains of previous crops, and both crop and weed seeds, allowing them to break down. It also provides a seed-free medium for planting an alternate crop. In modern use, a ploughed field is typically left to dry out, and is then harrowed before planting. Ploughing and cultivating a soil homogenizes and modifies the upper 12 to 25 cm of the soil to form a plough layer. In many soils, the majority of fine plant feeder roots can be found in the topsoil or plough layer.

Deep ploughing as done in Europe would be disastrous in tropical conditions since it would aggravate soil erosion. Ploughs were initially human powered, but the process became considerably more efficient once animals were pressed into service. The first animal powered ploughs were undoubtedly pulled by oxen, and later in many areas by horses (generally draught horses) and mules, although various other animals have been used for this purpose.

Figure 4-2: Horses drawn bottom plough

Sowing and planting

Work connected with planting and sowing differs very much according to the crops grown:

Selection and preparation of seed material: Selection of seeds, tubers, corms, suckers or cuttings as planting material always involves a lot of skill and careful observation. Each farming community has its set of rules in order to find out what will make the best planting material. Pupils could be asked to find out from their mothers and fathers how they recognize suitable planting material and what signs they look for when they reject a plantain sucker or a groundnut seed for planting. There are also a number of techniques of safe storage and of preparation of seed material for planting. One such technique is pre-germination (maize and bean seeds, seed yams) which advances plant growth after planting. These techniques should be described and discussed in school.

Actual planting: Crops are always planted by digging a hole and burying the seed material in the soil. Unlike in earlier European farming system, cereals are not broadcast but the individual seed grains are planted one by one. Some crops are planted together at the same time under the method of mixed cropping. Thus, maize and bean seeds or groundnut seeds may even be mixed in the same container used for planting. There are never less than three maize seeds put in a stand, often more, and the same number of bean or groundnut seeds.

Planting is not done in straight lines, or according to precisely measured distances. Since neither animal drawn implements nor engine-powered machines are used at any stage during farming, there is not really a need for straight lines. All that is required is that enough space is left for people to pass when they weed or harvest without damaging the crop. On a mound farmed with maize, beans, and leaf vegetables according to traditional methods, the average distance between stands was roughly 40 cm, with a standard deviation of 15.7 cm. This certainly does not represent a strict standard of planting distances. Most stands are between 25 cm and 55 cm apart from each other.

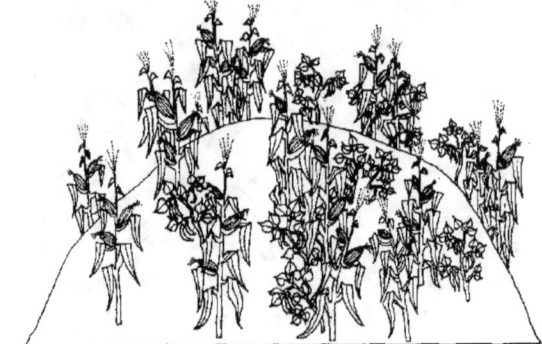

Figure 4-3: Mound planted with maize and beans (side elevation)

On the mounds, crop density is high and amounts to 6 - 7 stands per square meter. Making allowance for the paths between the mounds that use up quite a lot of land, this would amount to about 45 000 stands with at least two plants each per hectare. The study from Eastern Nigeria reports crop densities of 22 000 to 31 800 stands per hectare on compound farms immediately surrounding the house, and between 12 000 and 40000 stands on farms away from the compound (Lagemann, 1977)

Crop density varies a lot according to soil fertility, the crops grown, and the amount of preparatory work a farmer is willing to do. Well prepared soil will support a higher crop density than poorly tilled soil. When ridges are formed, as is normally the case across the slopes of hills, more or less continuous contour lines are formed. Ridges are of rather uniform width, depending on the work habits of the woman building them.

Planting on ridges is often done in staggered rows. For example, up to three rows of maize and beans or groundnuts or cowpeas may be planted on a ridge. Again, planting distances are not measured out by a yardstick, but they are not haphazard either. They follow rule-of-thumb knowledge about the best density on a given soil. If a school class went to measure the distances between stands of maize or cocoyams along one or two ridges planted according to traditional methods the children might be surprised by the degree of regularity they found.

Weeding

Weeding is a feature of nearly all farming. Exceptions are farms where crop associations are found which keep down most or all the weeds so that weeding becomes unnecessary. This is the case where pumpkins or various species of melons are grown as a secondary crop. Weeding is usually done with a hoe. Weeds are left on the farm to wither - except those that could immediately start to grow again. As they decompose, they add nutrients to the top soil. If they are available in sufficient quantities they act as mulch, protecting the top soil against erosion and loss of water. Weeding is very demanding in terms of labour. If it is done too late there will be serious damage to crop yields

The effect of weeding on plant development also depends on good timing. In traditional agriculture, weeding may be done later than in modern agriculture, due to the effect of fire on weeds:

". . . early weed growth after the burn was minimal, and as most farmers only weeded once, the one weeding gave superior results if performed later than very close to planting, when weed competition was minimal." (Lagemann, 1977)

Manuring

Under traditional agriculture, farms are usually not manured. As mentioned earlier, soil fertility is restored by either

1. Observing the fallow period, or
2. By burning at the beginning of a new farming season.

The cultivated areas directly surrounding the houses are manured with all sorts of suitable materials, e.g. kitchen and household refuse, ashes, animal dung, or waste from the processing of crops.

Compost is hardly ever prepared. The often-quoted study on traditional African farming systems reports the following methods of manuring the compound farm:

1. *Mulching*: all kinds of smaller branches, twigs and leaves from trees and shrubs are used for mulching the compound. The yam mounds especially are covered with a thick layer of these materials;
2. *Animal waste*: dung, mainly from goats, is applied throughout the year to the crops. It is usually applied to individual plants;
3. *Ashes*: as with animal waste, ashes are distributed directly around the individual plants;
4. *Composting*: This practice is common among the people of Owerri-Ebeiri (a village with a high population density) in Imo state, Nigeria where grasses from the fallow areas are collected and thrown together with household remains, into pits where the materials decay. The resulting rich soil is then applied to the crops;
5. *Shifting of latrines every year*: The latrines are shifted to another site within the compound. After a period of one cropping season the latrines are filled in, and bananas or plantains are planted, and so receive ample nutrients for many years.

These manuring habits have not been developed to such a degree in the lower populated areas, ... the labour-intensive way of building up soil fertility is a result of the people's efforts to overcome the increasing food shortage which results from the high population density." (Lagemann, 1977)

Where ridges are built, a certain form of green manuring is sometimes used: At the beginning of a new planting season weeds, grasses and crop residues such as maize stalks are cut and gathered in the furrows. After this, neighbouring ridges are divided up and new ridges are built in the former furrows. Crops growing on the new ridges not only profit from the decaying plant material buried in the soil but also from nutrients that rain may have washed down from the former ridges.

4.2.2 Mechanized farming activities in the early agricultural era

Modern (mechanized) methods of farming evolved out of the inherent shortcomings of traditional agriculture in meeting the demand of the evolution of forgers/hunter-gatherers into farming systems, rapidly increasing population, improved methods of farming, among several other reasons.

Various agricultural operations in which farm tractors and machineries are in common use include but not limited to:

1. Site selection and surveying,
2. Land clearing and forming,
3. Agricultural soil preparation (tillage),
4. Planting operations,
5. Post planting operations (crop establishment),
6. Harvest and material handling,
7. Crop processing and packaging,
8. Preservation and storage of agricultural products.

Site selection

Effective farm project or undertaken on mechanized scale started with site selection and survey. Site selection has to do with making a choice of land based on feasibility studies, soil conditions, topography, physical conditions etc. this can only be authenticated through land survey.

The importance of selecting a good site

1. **Future success**: The whole future success of your enterprise depends on the selection of a good site for your farm.

2. **Ease of farm management**: The layout and the management of your farm will largely be influenced by the kind of site you select.
3. **Control over construction cost**: The site therefore will strongly affect the cost of construction, the ease with which the farm can be managed, the amount of products produced and, in general, the economics of your enterprise.

Preliminary decision making

1. Before starting to look for a site, you should have a clear idea of the type of agricultural production you wish to build. Some of the questions you should ask yourself are the following.

    a. Which level of production do I plan to reach, subsistence or commercial? Which scale, if commercial?
    b. Which culture system shall I adopt?

    - Extensive or intensive;
    - One or several species;
    - Seasonal or year-round.

    c. Shall I use fertilizers or organic manures or both?
    d. Which type of crop or animal shall I produce and at which size shall I sell them?
    e. Shall I have to buy inputs or produce them all myself?
    f. Shall I try to integrate my farming activities with other agricultural productions? Shall I also start raising animals on my farm? One of the most important decisions when planning any livestock facility is site selection. The site for the feedlot operation must be suitable for housing, handling and feeding cattle.
    g. Which area of the farm do I wish to develop immediately? Will I develop other areas later, as a second phase?

2. If you cannot answer these questions by yourself, you should look for assistance, for example from the local extension agent specialized in specific area of farming of interest. You can also check with other farmers to find out what choices they made and why.

Land surveying

Webster's Dictionary defined surveying as "the science of determining the location, form or boundaries of a tract of land by measuring the lines and angles in accordance with the principles of geometry and trigonometry". This definition is rather outdated now with the advances made in geo-informatics.

Land survey is carried out for the determination of relative horizontal and vertical position, such as that used for the process of mapping and the establishment of marks to control construction or to indicate land boundaries. A simple survey of a building site provides accurate information needed to locate a building in relation to other structures or natural features. Data from the survey is then used for drawing a map of the site including contours and drainage lines if needed. Once located, the building foundation must be squared and leveled.

Land clearing development

Land clearing is the development of land with potential for agricultural use. Land clearing requires the removal of vegetation from the surface of land. This includes the removal of roots and embedded rocks. The land must then be broken in order to get a workable seedbed into which a crop can be seeded. Land breaking includes the removal of roots, stumps and rocks.

Clearing land to remove trees, shrubbery, rocks and debris is sometimes necessary, whether it be by a homeowner or a commercial developer. The reasons for doing so can range from making room for a vegetable garden to building an apartment complex.

Clearing brush

Clearing brush is often the first order of business when clearing land. For soft grasses and reeds, a sharp sickle mounted to a pole and used in a side-to-side motion will work sufficiently. For denser growth such as vines, a brush clearing machete is best. This variation on the standard machete is essentially a multi-tool with double-sided blades for slicing and hacking as well as a hooked end blade for especially thick foliage.

Removing trees and shrubs

Remove high-reaching shrubs as well as small tree branches with a lopper. A lopper utilizes two curved, scissor-like blades attached to long handles that provide increased leverage--sometimes up to six feet in length. Low lying shrub and tree branches can be cut with a smaller-handled lopper. A pruning saw can be used for especially thick branches. A bow saw can be used to cut trunks of very small trees and shrubs if complete removal is necessary. Regardless of what the scope of the job is, a variety of equipment is available to accomplish the task.

Hand tools: For the do-it-yourselfer who wants to clear a patch of land on his property, many of the tools required are those typically used for lawn work or gardening. Shrubs, bushes and small trees can be cleared by using pruners, loppers and hand saws to reduce size prior to using shovels to dig out the roots to completely remove them. Rakes, pickaxes, and hoes can be used for removing rocks and roots.

Bulldozers: Bulldozers are one of the most valuable and widely used pieces of equipment for large scale land-clearing needs. Most people visualize bulldozers with large, slightly curved blades that push dirt around. While this may be the most common means of clearing land, there are many attachments that fit on the front of the dozer in place of the blade that are designed to assist with even the most difficult land-clearing operations. Attachments include an implement called a root plough that utilizes a sharp horizontal blade that is pushed into the ground to remove all types of bushes and shrubs at the root level. Another commonly used attachment is a heavy-duty rake which is used to remove rocks, shrubs, small trees and stumps.

Excavators: Common in larger land clearing applications, these machines use a hinged boom on a rotating platform that operates much like a human arm. Attachments such as grapples, which function similarly to a human hand, enable the excavator to reach out, grab onto, and remove brush, small trees and logs to stack and load into trucks for disposal. Excavators also use other attachments such as stump pullers to cut lateral roots for easy removal of stumps, and heavy-duty rakes to gather large rocks into piles for loading into trucks.

The success of a project depends on favourable cost/benefit relationships, favourable environmental factors and ownership or management skills. A sufficient amount of information must be gathered and analyzed before undertaking any land clearing project.

Landform development

Land forming is the process of cutting, movement and distribution of soils evenly on undulating land to obtain level surface. This includes leveling in preparation for agricultural operation. The degree of land forming depends on the purpose to which the land is to be used. This depends on such factors as the depth of topsoil, original shape of the soil and the purpose to which the soil is to be suited. Landform activities involve operations carried out after initial land clearing. These include land forming, tillage and soil conservation.

Better land leveling results in improved crop yield, weed control, farm operation, seeding practices and efficiency of water use. *Other benefits and opportunities of land leveling include:*

1. Timely ploughing of the field,
2. Harvest of evenly ripened crops, and
3. More rapidly shedding of floodwaters from the field.

Ploughing

Ploughs advanced from a forked wooden stick making a furrow (often called a hog plough because they root in and out of the ground) to an iron tipped wooden plough initially pulled mostly by oxen. Depending on the type of soil, a typical team of 2-6 oxen and 1-4 men could plough about one acre/day. Many crops required ploughing up to three times in a season to prepare the fields for planting.

Harrowing

Harrows are typically pulled by draft animals or tractors and are used for breaking up and smoothing out the surface of the soil in preparation for planting. Some harrowing may be done to keep down weeds or as a type of low impact tillage. Harrowing is often done on fields to smooth the rough soil finish often left by ploughing. When planting by broadcasting seed the seed is often spread over harrowed land and then buried to about the right depth for growth by lightly harrowing the soil again.

Seed planting

Jethro Tull made early advancements in planting crops with his seed drill (1701) - a mechanical seeder that sowed seeds efficiently at the correct depth and spacing and then buried them so they could grow. Before the introduction of the seed drill, the common practice was to plant seeds by broadcasting (evenly throwing) them across the ground by hand on the prepared soil and then lightly harrowing the soil to bury the seeds to the correct depth. Other seeds were laboriously planted one by one using a hoe and or a shovel.

Figure 4-4: Jethro Tull's grain drill

Over 50 other inventors helped make his initial seed drill a machine in common use by 1900. It took a century and a half after Tull's publication, *Horse hoeing husbandry* in 1731 for farmers to widely adopt the technology. Although a pioneer in Europe, Tull was not the first to invent or use a seed drill; its origins can be traced back thousands of years to the East and China and other parts of Asia.

Figure 4-5: Broadcast barrow, courtesy of Roy Brigden

By 1850 the seed drill had competition with mechanical broadcast seeders with a crank that more evenly spray seed over the ground. Some broadcasters were combined with a wheelbarrow to allow more seed to be easily carried and distributed.

The advantages of the drill were that it was faster and it could be set up to plant seeds evenly to make weeding easier. The disadvantages were it was a new "expensive" machine that only worked well on particular soils and usually did not significantly increase crop yields. Today many seed drills use air pipes that allows the seeds to be blown from a bin to a group of disk "coulters" set at the desired row spacing that cut slots in the ground into which the regulated amount of seeds fall. Behind the coulters are spring tines, which help to cover the seed up.

Manure spreading

Manure spreaders used to distribute manure over a field as a fertilizer evolved from a simple farm cart to dedicated wagons or carts that was not used for other purposes. The first successful automated manure spreader was designed by Joseph Kemp in 1875. This and later manure spreaders used a drag chain to pull a board at one end of a manure filled wagon that forced the manure into a series of rotating tines that spread the manure more evenly. The power for the drive chain and rotating tines came from a connection to the rotating axle driven by the tires. They now have manure spreaders of several different types that work on dry, solid and slurry manures as well as irrigation systems.

Manure spreaders are still used today to help distribute this natural fertilizer. Application of waste products to agricultural land is an efficient method of recycling them, while at the same time improving the productive capacity of soils (Sommers, 1977) Manure still provides significant organic plant nutrients (fertilizer) to soil and significantly reduces the need for commercial fertilizers.

Weeding

Weeding progressed from a hoe or mattock powered by a man or woman to a hand pushed wheeled metal cultivators with steel or iron blades. The new development of the seed drill that allowed some crops to be planted in straight rows allowed further weeding advances. Cultivators were enlarged so they often straddled several crop rows and were pulled by draft animals or after about 1920 by petrol engine powered tractors whose wheels straddled the rows. For mass control of weeds harrows pulled by draft animals or tractors with cupped shaped steel discs and/or serrated rotating toothed discs were developed to chop down and cut the weeds up.

Herbicidal weed control

When herbicidal weed control was first widely commercialized in the 1950s and 1960s, it played into that era's optimistic worldview in which sciences such as chemistry would usher in a new age of modernity that would leave old-fashioned practices (such as weed control via cultivators) in the dustbin of history. Thus herbicidal weed control was widely adopted, and in some cases too heavily and hastily. In subsequent decades, people overcame this initial imbalance and came to realize that herbicidal weed control has limitations and externalities, and it must be managed intelligently. It is still widely used, and probably will continue to be indispensable to affordable food production worldwide for the foreseeable future; but its wise management includes seeking alternate methods, such as the traditional standby of mechanical cultivation, where practicable.

Haymaking

Hay is grass or legumes like clover and alfalfa or other herbaceous plants that have been cut, dried, and stored for use as animal fodder. It is typically used to feed grazing livestock such as cattle, horses, goats, and sheep during the winter or other times when food is scarce. The hay originally was cut with scythes, dried and collected with rakes and then put onto a wagon pulled by oxen or horses to a haystack where it was stacked for winter use.

The first major improvement seen in haying after about 1850s was when the sickle bar mowing head with its many triangular shaped knives mounted on a reciprocating long rod pulled by a team of horses was introduced. The cutting head extended to the side of the mower and the power for the reciprocating knives on the mower head was provided by the rotating wheels supporting the mower and driver. Horse drawn: mowers, rakes and haystackers were some of the first improvements seen in haying sometime in the mid-1850s and on.

Figure 4-6: Haymakers using scythes and rakes, *Grimani Breviary*, c. 1510[11]

By the late 1920s these machines were largely replaced by tractor drawn haying implements (many made originally for horses). Growing, cutting and storing the hay can be a long laborious process done two, three or more times per year.

Figure 4-7: Oxen powered hay wagon on the Gaisberg, near Salzburg, July 1903

Today hay is cut with wide hay mowers and left to dry. It is then raked into rows where is picked up by large balers that collect the cut and raked crop and then compress the hay into compact bales that are easy to handle, transport and store. The compressed bales are held together by baling wire or twine. The original balers made rectangular bales that were typically loaded, hauled and stacked into haystacks by hand. Today, some of the balers now make bales of hay that are rectangular or round and can weigh anywhere from 1,100 to 2,200 pounds (500 to 1,000 kg) each and have to be handled with powered equipment.

---

[11] c. 1510 implies about 1510

Figure 4-8: A Massey-Harris reaper-binder pulled by a tractor (Rutland, England, 2008)

Harvesting

Grain was originally harvested by cutting the grain with a sickle or scythe, tying the grain in bundles that were gathered into sheaves and allowed to dry before being brought to a harvesting barn by wagon where the grain was beat with hand powered flails to separate the grain kernels from its stem. The grain straw was collected and then hauled off for other uses. Harvesting was a very time consuming and laborious process that usually involved nearly all able bodied people on a farm or in a village for a period of several weeks.

Threshing

The thrashing machine, or, in modern spelling, threshing machine (or simply thresher), was a machine first invented by a Scottish mechanical engineer Andrew Meikle for use in harvesting grains. It was invented (in c.1784) for the separation of grain from stalks and husks (straw). Many other farmers and inventors improved the original design over a period of many years. Power for the thresher came initially from a team of about six to eight horses walking in a circle around a bullwheel, which was connected by long belt, which powered all the different thresher parts. The original thresher was stationary with stemmed grain fed to it manually and the separated grain and straw removed manually. Figure 4-9 shows a typical threshing machine powered by two horses in 1881.

The separated grain was usually put in sacks that were sown together by two men who sewed up to 1,000 sacks a day. Other workers were needed to drive wagons to load, deliver and unload the grain to where it was stored and haul off the straw. A harvest crew could be as large as 18 to 30 people with 20 to 30 horses used to power all the wagons and the thresher.

Figure 4-9: Threshing machine, 1881

One of the first improvements made was to replace the bullwheel and its teams of horses with a stationary steam engine turning a large pulley and a long wide leather belt(s) to power the thresher and its various moving parts. Steam tractors were often made to be primarily stationary steam engines needed to power threshers and used coal, wood or straw to make steam. Steam engines were also used to power threshing machines. There were steam engines that moved around on wheels under their own power for supplying temporary power to stationary threshing machines. These were called *road engines*, today both reaping and threshing are done with a combine harvester.

Threshing mechanism

The cut grain and stems (straw) were originally fed by hand into the thresher, which was stationary as the grain with its heads, stems and chaff was brought to it. The threshing mechanism consisted of a rotating threshing drum (commonly called the "cylinder"), to which grooved steel or hard rubber bars (rasp bars) are bolted. These rotating bars pulled the grain through the thresher.

The rasp bars thresh or separate the separate grain kernels in a grain head and the chaff from the stem through the action of the pushing and bending the grain head against the steel meshed grilled "half cylinder", or *concave*, through which grain kernels, chaff and smaller debris falls as they break off the stem. The straw or grain stem, being too long and light, is carried through the concave onto the straw walkers to be conveyed out of the combine at a separate outlet. A separate blower section in the thresher blows the lighter chaff off the grain to get a chaff free grain. The grain is heavier than the straw and chaff, which causes it to fall rather than "float" across from the cylinder/concave to the straw walkers.

These machines were typically so large that it may take a team of 18 to 30 horses to move them from one harvest field to the next. The large harvest crew of 18 to 30 men typically stayed together for the whole season working 10-12 hour days seven days a week till the harvest was done. The cut grain, chaff and straw typically raised large clouds of dust the harvesting crews had to work in. Usually the women on the farm or adjacent farms being harvested by that crew prepared the meals for the men on the harvesting crew. Additional help was usually traded with other farmers and their families.

Winnowing

The next operation was to separate the chaff from the grain kernels by throwing it into the air in a wind and letting the heavier grains fall to the floor while the chaff blew away. The grain was then typically sacked for storage or sale. Wagons were again used to haul off the grain and straw.

Combine harvesters

A major improvement was to marry the reaper and the thresher in a single machine called a combine. In the 1880s the reaper and threshing machine were combined into the combine harvester. The combine name derives from the fact that it combines three separate operations, reaping, threshing, and winnowing, into a single process. These machines required large teams of horses or mules to pull. Steam power was applied to threshing machines in the late 19th century.

The first combines were made in the 1850s but it took till after the 1920s for them to become common (Constable et al., 2003). Advertising for motorized equipment in farm journals during this era did its best to compete against horse-drawn methods with economic arguments, extolling common themes such:

- A tractor eats only when it works,
- That one tractor could replace many horses, and
- That mechanization could allow one man to get more work done per day than he ever had before.

In the original versions the bullwheel and its teams of horses that powered the thresher was replaced with large steel drive wheels that turned as the whole combine apparatus was pulled across a grain field with large teams of up 24 to 30 horses. The wheel's outer surface turned the cutting bar and provided power for the thresher as the implement was pulled forward and the support wheels rotated. The swath bar with its sickle bar cutting head in the front of the combine cut the grain as it progressed. The large rotating reel above the cutter bar pushes the grain into the sickle bar and then pushes the cut grain onto the combine's first conveyor belt that delivered the grain to the threshing part of the combine.

Soon power for the various mechanisms in the combine was supplied by small steam engines mounted on the thresher and the power connection to the drive wheels was eliminated. These engines were replaced in turn by petrol engines as they became available after about 1900. As tractors became available they replaced the large team of horses used to pull the combine.

These types of pulled combines with petrol engine powered threshing mechanisms were used until the 1940s in some areas. One of the disadvantages found in pulling these often massive, top-heavy machines on hilly fields was the risk of overturning — an expensive and dangerous event. This was partially solved by constructing combines that could *lean* into the hill while their harvesting head was tilted to follow the sloping ground.

Many small farmers could not afford to buy these expensive, complicated machines and either hired someone to custom cut their grain for a fee or used what "old" equipment they had. Combines increased the harvest speed, cut the crew needed to harvest the crops from over 30 workers down to today's two to three much more productive workers and made grain "cheap" enough to help feed the world.

Modern combines

At first reapers and combine harvesters were pulled by tractors, but in the 1930s self powered combines were developed. The combine harvester of today, or simply **combine**, is a self-propelled machine typically powered by up to a 400 hp diesel engine. Among the crops harvested with a combine are wheat, oats, rye, barley, corn (maize), rice, soybeans, flax (linseed) and other grains. The grain is cut by removable cutting heads 20 feet (6.1 m) to 40 feet (12 m) wide designed for specific crops.

Figure 4-10: Self-propelled Claas Lexion 570 combines

Figure 4-10 shows a self-propelled Claas Lexion 570 combine with enclosed, air-conditioned cab with rotary thresher and laser-guided hydraulic steering harvesting oats. The cut grain and stem is typically transported to the threshing mechanism in the combine by various conveyor belts. The separated grain is typically stored in a bin mounted on the combine. The bin has a movable auger system to empty its contents into an adjacent truck or wagon. Often the bins are emptied while the truck or wagon matches speeds and the combine continues its harvesting.

The straw typically left behind on the field is the dried stems and leaves of the crop with the grain removed. This "straw" is either spread on the field by the combine for future use as organic fertilizer or raked and baled for livestock feed and bedding. The high cost (over USD $400,000 for some) of these combines has led to "custom" cutting where a combine or several combines are bought and run by an individual or company to cut many different farmer's fields for a fee. Custom cutting allows the cost of the combine to be spread over many farmers as a combine can often run over several weeks or months of operation (Constable et al., 2003). A combine with a 40 feet (12 m) cutting bar can harvest 80–200 acres (32–80 hectares) per 12–15 hour day and may be able to be run daily for nearly two months.

Part 2

AGRARIAN MECHANIZATION:
THE REVOLUTION

Horse-power transition to machine-based agriculture

CHAPTER 5

MECHANIZATION
The Historic Developments

5.0 Introduction

Throughout most of its long history, agriculture — particularly the growing and processing of crops — was largely dependent on human sweat and later, on draft animal labour. Even as late as the 19th century, farming and hard labour still remained virtually synonymous; hard and hazardous, and productivity had not shifted much across the centuries.

It is interesting to know that draft animals such as oxen, horses, and mules were earlier employed to pull ploughs to prepare the soil for seed planting and were also extensively used to haul wagons filled with the harvest — however, it is more appalling that up to 20 percent of the harvest often went to feeding requirements for the animals themselves. The rest of the chores required backbreaking manual labour: planting the seed; tilling, or cultivating, to keep down weeds; and ultimately reaping the harvest. All these processes are complex and arduous tasks requiring cutting, collecting, bundling, threshing, and loading.

However, the arrival of motorization (mechanized agriculture), from the late nineteenth century and the beginning of the twentieth century, has changed agricultural production and the whole economy by dramatically reducing the work force and improving productivity. Farm mechanization and other technological advances such as fertilizer and agro-chemicals, irrigation and drainage etc had made possible increase in productivity to such an extent that in United States of America for instance, in 1910, a farmer could only supply food for about 8 people; and 80 years later in 1990, he or she could supply food to as many as 90 people (White, 2000). And in this productivity growth, tractor, one of the most important innovations of the farming network, played a fundamental role.

Limiting factors of traditional agriculture

As earlier pointed out, the rise of agriculture had changed significantly the world outlook. Long before the green revolution paved the way for vastly improved yields, people were notoriously bad at using the land. For instance, to produce our food, we used to cut down a staggering number of trees and, without inputs like fertilizer or irrigation, or the massive intertwined agricultural system we have today – fire, drought, and flood – had cut vital food supplies for many years. So, while the rise of agriculture allowed human populations to blossom, it also opened the door for catastrophic cultural and environmental collapses. Attempts at proffering solutions to these arduous problems led to what is today known as mechanization.

5.1 Mechanized agriculture

Mechanization is the process of developing agricultural machines and technology and substituting this machine power for human and animal power in agricultural production practices to greatly increase farm workers' productivity. Mechanization improves production efficiency, quality of produce, and encourages large scale production. In its simplest term, mechanization is an increase in production per worker per hectare of land cultivated. This had been made possible at three common fronts; mechanized agriculture through; 1) machinery development, 2) water conservation through pump and irrigation development and 3) soil conservation through conservation tillage.

Agricultural machinery development

Mechanized agriculture is largely characterized by machines or equipment powered or attached to mechanical power sources (i.e. steam engines, gas turbine, gas engine, gasoline/diesel engines, or any other form of prime mover) for the purpose of work delivery.

In the 1850's, the industrial revolution spilled over to the farm with new mechanized practices which grossly increased production rates. Earlier on, before 1850, there had been significant changes witnessed in the development and use of early farm implements. Most of these early implements were then and still powered by horses or oxen. These new implements combined with crop rotation, manuring and better soil preparation lead to a steady increase in crop yield.

The advent of steam power and later gas powered engines brought a whole new dimension to the production of crops. Yet, even as recent as 100 years ago, four-fifth ($4/5^{th}$) of the world populations lived outside towns and was in some way dependant on agriculture.

Water supply and irrigation development

The effect of water needs on the mechanization of agriculture has been profound, especially on-farm water-use. On-farm water use can be reduced substantially without decreasing productivity through improved irrigation technologies and efficient water management practices. Irrigation, described as the application of water to land using means other than the natural rain, for the purpose of providing sufficient water for plant growth and productivity provides one of the greatest possibilities for increased production.

For instance, at the beginning of the 20th century, only about 16 million acres of land in the United States were irrigated. This number had staggeringly increased, typically by intricate networks of gated channels that fed water down crop rows.

Before the demonstration of the first hand dug irrigation well by John J. Thieezene in 1931, most farmers depended almost exclusively on rain falling directly on their fields to meet plant-water supply requirements. Early Egyptians practices of irrigation used large basins of water from the Nile River to flood and irrigate their fields. The Mesopotamians utilized large basins of water and canals from the Tigris and Euphrates Rivers. Armenians used underground tunnels to convey water from springs in the foothills to their crops. These methods were often inefficient and caused crop damage.

Through the centuries farmers used these ancient ways of irrigation or depended on the unpredictable rains to grow and produce crops. This included farmers who would eventually settle in the American Midwest called the Great American Desert by many. This farmland required a lot of water.

From the foregoing, it is evident that irrigation has been around for thousands of years, for as long as people have been growing crops; however with low efficiency. Irrigation therefore needed a change... a revolution in increased water supply and high efficiency performance systems. Thieezene's F30 tractor (apparatus) running full speed provided the first solution which could be used to produce around 1,000 gallons of water per minute. His model of early well-drilling apparatus shown to the right of Figure 5-1 was used to haul soil up from the bottom of a hand-dug well (up to 140 feet inside a 40-inch wide hole). His invention paved way for the manufacture of today's drill rigs.

The desired change in increased productivity and efficient water supply needs was brought about by a tenant farmer and inventor from eastern Colorado named Frank Zybach in the 1940s who invented the center pivot system. Before the center pivot system, field irrigation was achieved by either of the following methods;

Figure 5-1: Thieezene set up with his son Albert and Walter Ott.

Ditch irrigation: Ditch irrigation worked by pumping water down furrows in-between the rows of crops. This required the field to be perfectly level or slightly sloped one direction and requires over watering to one end of the field. These ditch systems were constructed with horse-drawn plows and men moving rocks by hand and pick and shovel. With a lot of sweat and hard work, they turned many a dry sagebrush flat into a productive green hayfield over time.

Water wheel: The water wheel, used for generating power, was also utilized for irrigation. This was done by running water under a wheel. The turning wheel picked up water and dumped it into an elevated canal. This required fields to be near a source of running water.

Gated pipe irrigation: Gated pipe uses pipes with small gates covering holes in its sides. When water is sent through the pipe the open gates allow water to pour out between the rows. While more efficient than ditch irrigation, water is still lost into the walls of the rows.

Siphon tubes: Siphon tubes work by using suction to convey water out of a concrete or soil ditch and into the furrows between crops. Siphon tubes have to be moved from field to field and require lots of work to get the flow of water started. [Siphon tube irrigation] uses quite a lot of water—far more than a typical center pivot system.

Sprinkler irrigation: Sprinkler irrigation is done by pushing water out through pipes which create a pressurized spray. This can only be used on certain crops due to their height and it requires a lot of sprinklers.

Then in the 1940s Frank Zybach devised something better— a self-propelled sprinkling irrigating apparatus later to be known as center-pivot system, which consists of sprinklers attached to a pipe that runs from a hub out to a motorized tower on wheels.

As the tower moves, the sprinkler pipe rotates around the hub, irrigating the field in a grand circular sweep, now known as center pivot irrigation. Zybach applied for a patent on his invention in July 1949 and was granted patent in 1952 as the self-propelled sprinkling irrigating apparatus. Along with other mechanized systems, it has almost quadrupled irrigated acreage and has also been used to apply both fertilizers (fertigation) and pesticides (chemigation).

Figure 5-2: Frank Zybach apparatus (left) near Strasburg, Colorado

The July 22, 1952 patent of the self-propelled sprinkling irrigating apparatus marked the birth of a revolution in irrigation technology. The center pivot, in less than fifty years, has revolutionized irrigation as a part of agricultural production.

Soil and water conservation development

Deficiency of soil moisture for crop usage is another critical aspect of agriculture namely, soil and water conservation. These problems are most serious essentially in the arid/desert regions where rainfall distribution is not uniform in the growing season. To solve these perennial problems, the agricultural engineers and other scientists have humorous duty in developing new practices that will permit storage of greater percentage of the available precipitation in the soil profile.

Mechanization has come to the aid of this critical aspect of agriculture with an approach known as conservation tillage which has greatly reduced, or even eliminated, traditional ploughing, which can cause soil erosion and loss of nutrients and precious moisture. Conservation tillage includes the use of sweep ploughs, which undercut stubble but leave it in place above ground to help restrict soil erosion by wind and to conserve moisture. The till plant system is another conservation-oriented approach. In this approach, corn stalks are left in place to reduce erosion and loss of moisture and at planting time the next year the row is opened up, the seeds are planted, and the stalks are turned over beside the row, to be covered up by cultivation. This helps conserve farmland by feeding nutrients back into the soil.

Driving force of agricultural mechanization revolution

At the turn of the 20th century the introduction of the internal combustion engine set the stage for dramatic changes to be witnessed in that centenary agricultural revolution. Right at the center stage of agricultural mechanization revolution was the tractor.

Tractors pulled ploughs, hauled loads and livestock, towed and powered the new planters, cultivators, reapers, pickers, threshers, combine harvesters, mowers, and balers that farm equipment companies kept coming out with every season.

In all these, farming and farm machinery have continued to evolve. The hand held tools and apparatus had gave way for animal drawn equipments, ploughs are not used nearly as extensively as before, the disk harrow today is more often used after harvesting to cut up the grain stubble left in the field while the seed drills are still used, the air seeder is becoming more popular with farmers to cultivate many more acres of land than the machines of yesterday, the grain binder has been replaced by the swather which cuts the grain and lays it on the ground in windrows, and the threshing machine has given way to the combines with more sophisticated automation systems.

The following sections extensively discussed the evolutionary trends in agricultural machinery development history from the primitive agricultural era as well as the subsequent machinery revolution that characterized the 20th century.

5.2   Historic tool development in primitive agricultural era

The Neolithic period (primitive agricultural era) saw extensive development of tools for agriculture and mining activities. An array of Neolithic artifacts, including bracelets, axe heads, chisels, and polishing tools is shown in Figure 5-5 below. Rough-out tools were made locally near the quarries, and some were polished locally to give a fine finish. This step not only increased the mechanical strength of the axe, but also made penetration of wood easier. Flint was still used from sources such as Grimes Graves but from many other mines across Europe.

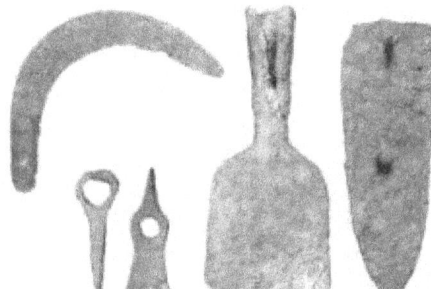

Figure 5-5: An array of Neolithic artifacts

For instance the stone axes were being made from about 3000 BC not just from flint, but from a wide variety of hard rocks from across Britain and North America as well. They include the noted Langdale axe industry in the English Lake District, quarries developed at Penmaenmawr in North Wales and numerous other locations.

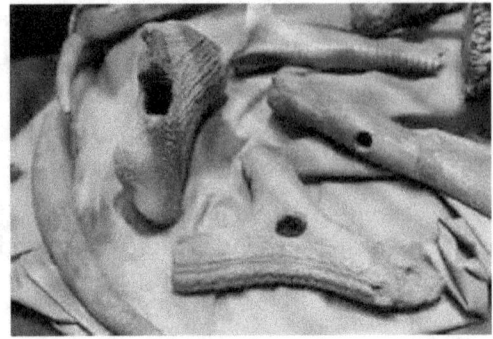

Figure 5-6: Neolithic tools cut from anther bones

Hoes: When people first began to work on the fields, they used their own power and hands to till the soil using a hoe to cut down weeds and prepare the soil for planting; an axe to cut down trees and shrubs or fire to help clear the forest for planting. Hoes are tools with metal blade-shaped end with the lower cutting edge, used to move earth. Hoes with blades of stone or bone were also widespread, and were mostly used for weeding and for forming soil into mounded furrows for ease of irrigation. An old Asian hoe and typical garden hoe could easily remind us of the timelessness of these ancient tools in gardening and horticultural practices.

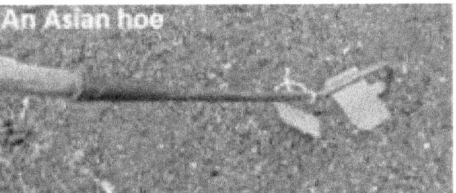

Figure 5-7: Old hand hoes

Digging sticks: Digging sticks are long, pointed shafts carved out of wood, often with a stone weight bound to the shaft near the digging point to add heft (weight) and balance. They were used to break up the soil and dig holes or furrows for planting seeds.

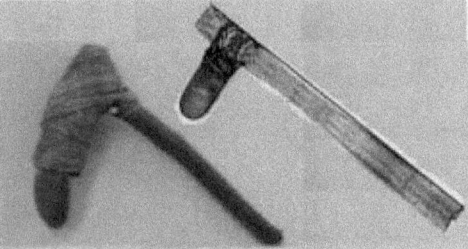

Figure 5-8: Digging sticks

Digging sticks and hoes remained the primary tools of cultivation until the advent of the plow during the Bronze Age.

"Ard" ploughs: As earlier pointed out, the first agricultural tools used in field operations were probably the hoe because of its primitive design, followed by the foot "ard" ploughs, which are more advanced in design, then the inventions of the Mediterranean "scratch" or swing plough (Figure 5-9).

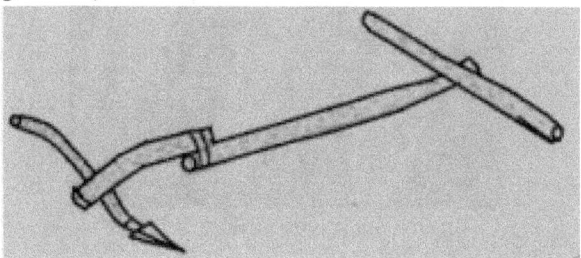

Figure 5-9: The scratch plough invented around 6,000 BC

The swing plough being more advanced in design, has no wheels to allow the ploughman control the depth and width of the furrow as he work. The scratch plough, essentially comprises of a sharpened (shaped) stick with handles for guidance and a pole for attachment to an animal or a human, was adapted to light or coarse textured soils.

Axes: The polished stone axe was one of the most important tools in Stone Age agriculture. It consisted of a flint or obsidian head bound to a wooden shaft, chipped into a cutting edge and then polished to a smooth finish. A polished axe can cut more deeply and easily into timber than earlier unpolished examples, allowing Neolithic farmers to quickly clear large areas of farmland for cultivation. In most parts of the world, early village-scale agriculture would not have been possible without an effective stone axe.

Figure 5-10: Stone axe

Blades: Blades are sheets of metal, preferably steel, used to till the earth which can be pointed or wide, with long cutting edge and wooden handle. Typical examples include knives, machete etc. Machetes are designed to cut, have a long steel blade sharpened as cutting edge.

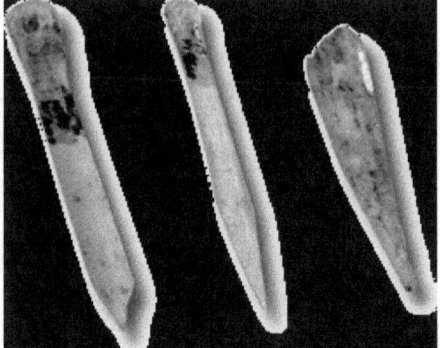
Figure 5-11: Neolithic machetes

Spud: Spud are primitive agricultural instruments composed of a steel part whose ends terminate in the form of rectangular blade, on one hand, and land vertically, has a rectangular blade edge and lower edge of wood or metal handle.

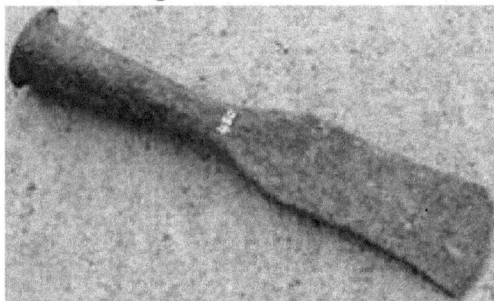
Figure 5-12: Iron weed spud

Sickles and harvesting sticks: Once a crop grown has reached maturity, it had to be harvested. Notable ancient harvest tools for harvest include the sickle. Some grains, such as rice, could be harvested by knocking the grain off the stalk with a short, heavy harvesting stick.

Figure 5-13: Harvesting sticks

Harvesting sticks often featured stone weights similar to those found on digging sticks. Other crops had more tightly-bound grains, and had to be harvested by cutting the grain-bearing head away from the stalk.

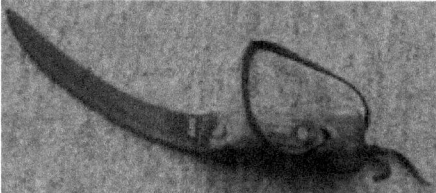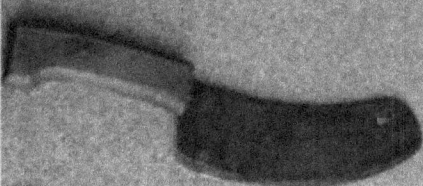

Figure 5-14: Left: Threshing bond culter, Right: Root crop toping knife

Wheat and barley were harvested with stone knives or sickles. A Stone Age sickle consisted of a broad flint or obsidian blade attached to a wooden haft, and resembled a narrow axe. Sickles allowed grain to be harvested closer to the ground, providing straw for animal fodder as well as grain for human consumption.

Figure 5-15: Wooden sickles

Grinding stones: Most Neolithic crops were cereals that had to be ground into flour or meal before they could be consumed. Many hulled grains pass through the digestive system whole, providing few nutrients unless they are boiled or ground. A grinding stone consists of two samples of the hardest stone available, usually dense metamorphic rock such as basalt.

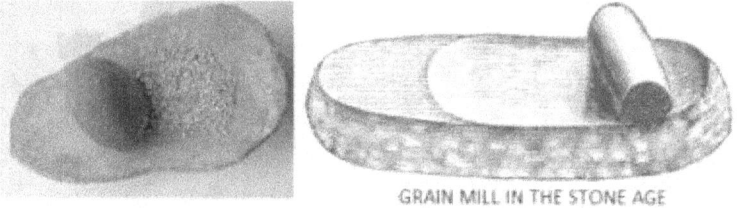

Figure 5-16: Grinding stones

A small hand grinder is used to pulverize grain against a larger pounding surface, which was sometimes dished to hold more grain without spilling. Once the grain was thoroughly ground, it would be brushed into a basket or pot for consumption. Grinding grain was a daily task in most Neolithic societies, since flour is harder to store than whole grains. Stone-ground flour was better than flour ground by hand.

Handling tools: These are wide blade tools with long handle used in handling bulk materials such as soil, rice, cowpea etc.

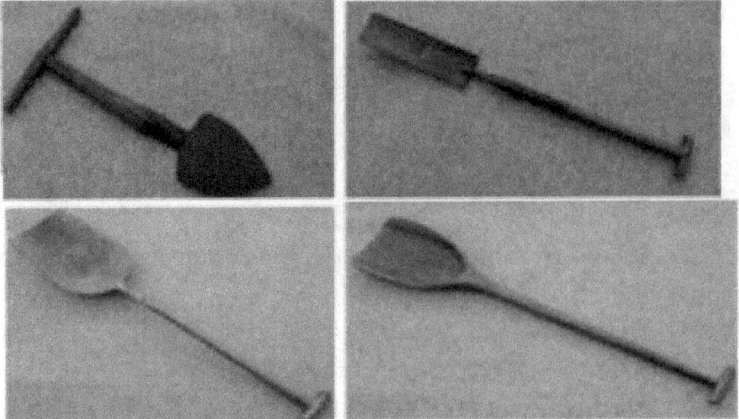

Figure 5-17: Handling tools: spades

Rakes: Rakes are ancient tools designed to gather or rake seeds and rubbish or refuse littering the ground. It has a horizontal metal portion forming thin or thick teeth depending on usage, and a long wooden handle.

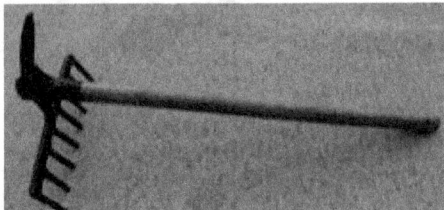

Figure 5-18: Stone rake

Transplanting tools: Spoon-shaped small metal blades, with sharp edges and wooden handle served to remove seeds.

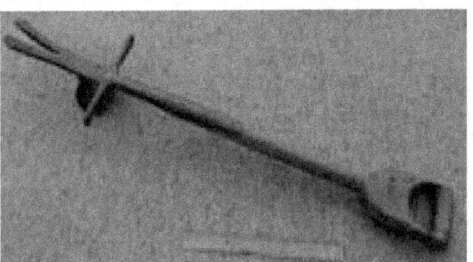 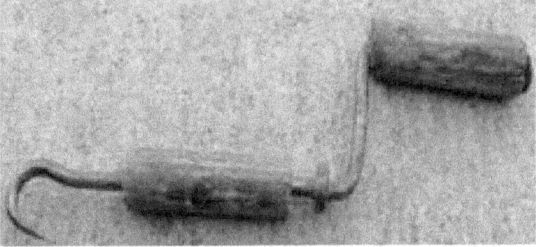

Figure 5-19: Left: Gore Snud, Right: Wimble

Watering can: Watering cans are portable metal containers with water tank with a spout ending in a round room with many small holes, used to water plants. The term watering can first appeared in 1692 in the dairy of keen cottage gardener Lord Timothy Simon George of Cornwall. Before then, it was known as a watering pot. In 1885, the "haws" watering can was patented by Michael Deas, replacing the top mounted handle with a single round handle at the rear and has since been improved.

Plastic watering can

Figure 5-20: Watering cans

Primitive agricultural tools are crude and inefficient, not capable of meeting the needs of food supply for the rapidly growing and transforming Neolithic society into the modern agrarian society with increasing human population as well as the emerging industrial revolution. These needs and development gave rise to the traditional agricultural era.

5.3 Historic tools development in traditional agricultural era

Traditional agricultural tools are as old as the Old Stone Age. These tools are economical in terms of labour, and are money and time saving also. They were made from locally available materials like; wood, iron etc. Traditional tools are operated easily without any special skills.

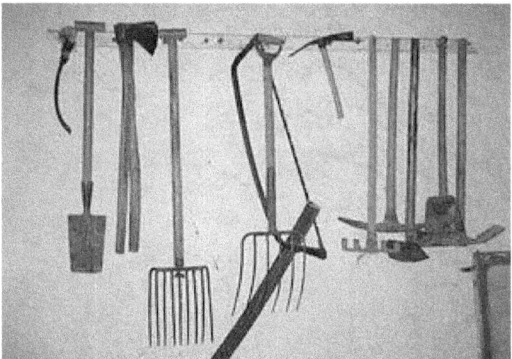

Figure 5-21: Agricultural hand tools

Agricultural tools are many and varied, among mentioned are: shovels, axes, hoes, mattocks, rakes, pitchforks, scythes, cradles, flails, wooden ploughs, oxbows, chains, knives, scissors, saws, hammers etc are classified as manual handling tools.

There are no significant developments that distinguished the traditional agricultural era from the primitive tool development except the change in materials of construction from the stone to wooden tools as well as the introduction of iron tools. Power supply still remains predominantly human with gradual domestication and utilization of animal drawn systems which ushered in the mechanized agricultural era.

5.4 Historic power development in mechanized agricultural era

Human power

This is the most basic level of agricultural mechanization, where a human being is the power source, using simple tools and implements such as hoes, machetes, sickles, wooden diggers, etc. Human power accounts for the lion's share of work in overall agricultural production from the primitive agricultural era.

Figure 5-22: Threshing rice manually

The amount of human power-use and intensity has been a source of concern to agricultural engineers owing to low level of development and natural limitations of human powered tools and machines (Mrema and Mrema, 1993; Comsec 1990; Anazodo *et al.*, 1989; and Anazodo 1987).

Figure 5-23: Manual milling with mortal

Their predominance in the agriculture of developing countries is an important factor to address when dealing with overall economic development of those countries. The problem of limited cultivated area of land is not necessarily with the tools used, especially for primary production operations, since efforts have been made to redesign them but this yielded no significant improvements (Makanjuola, 1991 and Odigboh, 1991).

Animal power

When human power was not enough to meet the needs of man, they used animal power, and then the power of machines, to make their work easier by first broken up the land with a plough and then, a harrow pulled by draught (draft) oxen or bullocks and horses which originally, were the major means of multiplying the efficiency of farm labour.

The ox power: The first animal power to be domesticated and widely applied to farming in Britain was the ox (a castrated bull (steer) above four years old) employed to pull ploughs, harrows, carts and wagons. Although a large ox is nearly as powerful as a large horse and had the advantage that they could be sold for food at the end of their lives, they were gradually replaced by horses. Oxen could also survive on poorer feed and unlike horses did not need supplements of grain for long work periods. Despite their advantages as draft animals, cows after about 1800 were primarily bred for either beef or dairy production and seldom as draft animals.

Figure 5-24: Teams of oxen pulling stumps

Ox harnesses: Ox harnesses development is a major innovation in the animal domestication and utilization history. These harnesses made animal control easier on the field. They include the straight oxen yoke, a mechanism for attaching drawn implements to the animal; it was the most common yokes used for oxen in the 19th century. This small yoke have been used to train and control a team of oxen to pull farm implements and vehicles. It comprises of a top wooden bar which fits on the back of the oxen's neck and the U-shaped wood bows slide up under the animal's neck. The bows are held to the bar with wood or iron pins.

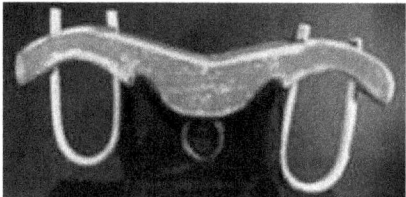
Figure 5-25: Oxen yoke c.1800-1860

Ox shod is another type of harness. Oxen could be shod also but this was a much more complicated process since cows have trouble standing on three feet and has cloven hoofs. As the farmers switched from ox power to horse power, most of the harnesses were redesigned to fit the horse while the drawn tools could almost be carried in a small cart or wagon. Though strong and durable, oxen did not take well to hard surfaced roads and were about 10% slower compare to horses, so were replaced by horses which were faster and often easier to work. Starting at about

The horse power: The ox power was gradually displaced after 1700 and by 1750; horses like the Shire horse and Clydesdale horse had replaced oxen as draft animals. The most popular draft horses in Britain at that time, which were originally bred to carry fully armoured knights, were the Shire horse and Clydesdale horse. Shire horses are one of the largest horse breeds

Figure 5-26: Shire horse

Horses were primarily used as pack animals and for riding. In South Australia, the English Shire or English Black Horse [12] and lighter Clydesdale[13] draught horses were most generally used for farm work, although many crossbreds and hacks also contributed to farm transportation and work associated with wheat production.

Figure 5-27: Clydesdale horses pulling spike harrows, Murrurundi, NSW Australia

---

[12] The horse described as the, 'English Black Horse,' is most prevalent, being in some degree connected with most of the common kinds of cart-horse. They are characterized by short, thick, ungainly body; strong, thick, and hairy legs, from the knee downwards; slow in action and heavy, useful only for slow work.

[13] The Clydesdale horse may be placed intermediately between the heaviest and lightest draught horses, combining as he does the very valuable qualities of power and free action.

Horses were used to power machinery as early as the 16th century; there are evidences of horse drawn stationary sweeps illustrated in Germany, by Agricola in 1556 and described by Ramelli in 1588, but not extensively used until 1700s.

Horse harnesses: An event that aided in the exploitation of the horse was the development of the lowly horse shoe and nails (Figure 5-28). This obviously helped greatly and permitted field work to be done under a wider range of soil and weather conditions, since the shoe gave greater traction and helped prevent hoof rot. Horses could be easily shod with iron horseshoes to protect their feet on hard surfaces. The horseshoes would typically wear out in 6–8 weeks and have to be replaced.

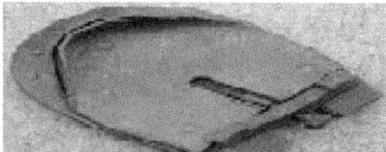
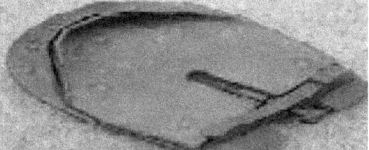

Figure 5-28: Horse shoes

Also of equal importance was the development of the horse collar. A collar similar to one used on camels was introduced from the east, perhaps from Bactria[14], reaching Europe around 800 or 900 A.D. This device permitted the exploitation of the horse. It is hard to believe that neither the Greeks nor Romans (who represented the "Classic period" for historians) were able to fully exploit the horse because only a variation of the oxen yoke was originally used on the horse.

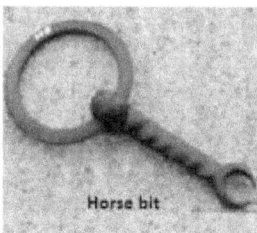

Figure 5-29: Horse collar and bit

Horseshoes, collars, bits and harnesses used by farmer in the 1500 to 1700s were made mostly of wood with the farmer creating them from the raw materials available. Horses became literally harnessed to the ploughs (Figure 5-30) and wagons when the horse collar was widely adopted for use in Britain after about 1700.

---

[14] Ancient Bactria was in present-day northern Afghanistan, between the Hindu Kush mountain range and the Amu Darya.

Timeline of agricultural mechanization

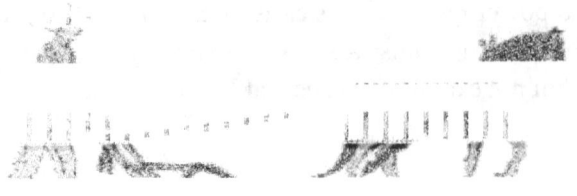

Figure 5-30: Plough horse with collar

A plough beam or wagon tongue would have been attached to the yoke through the iron ring which hangs down between the neck bows (Figure 5-31). The light harness for a single horse between shafts shown in horse 1 was used with two-wheeled and lighter four wheeled vans and wagons. The heavy type for a single horse between shafts was used for most four wheeled wagons and heavier four wheeled vans.

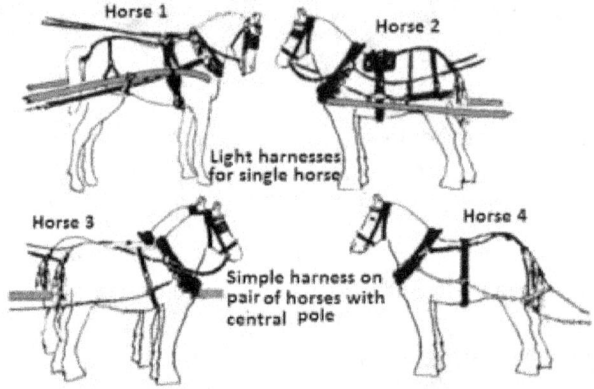

Figure 5-31: Attaching horse harnesses

The pair of horses with the central pole shown on horse 3 (Figure 5-31) illustrates the simple harness employed with this arrangement. Note how the reins pass through loops on the harness to ensure they pull down when used. Horse 4 in the lower right corner of Figure 5-31 is rigged for pulling a heavy load, perhaps some logs being dragged along on the ground. The same harness was used when pulling railway wagons about, the ends of the chains being fitted with hooks for this purpose.

In all cases the actual load is taken by the collar round the horses' neck, the remainder of the harness is there to support the pole or shafts and to guide the reins so they act in the desired way (pulling the horses head down and back).

Horse brass: Another horse harness is the horse brass[15]. A horse brass is a brass plaque used for the decoration of horse harness gear, especially for shire and parade horses. Common on horses hauling a brewers dray they would generally only be seen on farm horse harnesses at shows and when attending a market. They became especially popular in England from the mid-19th century until their general decline alongside the use of the heavy horse, and remain a collector's item today.

In the ancient Rome, horse harnesses were sometimes embellished with horse brasses, known as *phalerae*, cut or cast in the shape of a boss, disk, or crescent, most often used in pairs on a harness. In the medieval England, these decorative brasses were in use before the 12th century, serving as talismans[16] and status symbols.

Figure 5-32: Horse brasses

There are great deals of die-hard, unfounded myths surrounding these decorations such as their usage as amulets such as those of Charlemagne[17] to ward off the "evil eye". However, extensive, original research by members of the National Horse Brass Society has shown that there is no connection whatsoever between these bronze amulets to the working-class harness decorations used in the mid-19th century which developed as part of a general flowering of the decorative arts following the Great Exhibition of 1851.

---

[15] 'Horse Brasses' are cast brass ornaments which are attached to horses harnesses, they were originally 'magic charms' intended to ward off the 'evil eye', by the time the railways arrived these were (mostly) just used for ornament.

[16] A talisman is an object which is believed to contain certain magical or sacramental properties which would provide good luck for the possessor or possibly offer protection from evil or harm.

[17] Charlemagne (2 April 742 – 28 January 814), also known as Charles the Great or Charles I, was the King of the Franks from 768, the King of Italy from 774, and from 800 the first emperor in western Europe since the collapse of the Western Roman Empire three centuries earlier.

CHAPTER 6

AGRICULTURAL TRACTORS
Historic developments

6 Introduction

The arrival of motorization, before the beginning of the 20th century, has changed world agricultural production outlook and economy by dramatically reducing human work force and improving productivity. Yet the negligence of research in the area of farm mechanization history is surprising given that the history of economic growth has been strongly affected by farm mechanization. The history of the early traction engines and the mechanism behind the birth of other new technologies that drives agricultural transformation were equally explored. In this chapter, the developments that gave rise to the evolution of a major agricultural input called the farm tractor are investigated.

6.1 Historic development of agricultural engines

Replacement of animal traction by mechanical powers

As early as the 16th century, there are evidences that horses had been used to power machinery. For example, drawings of horse-drawn stationary sweeps were seen illustrated in Germany, by Agricola as early as 1556 and described by Ramelli in 1588, but not extensively used until 1700s when they were fully developed. Apart from the 19th century horse power development, another key element in the industrial and agricultural revolutions of late 19th and early 20th century was in the application of wind, water and steam power to drive machines. These developments are so significant to agricultural mechanization and development that their historic review had become imperative.

Wind and water power development

Windmills: Windmills were developed in the Middle East and only arrived in Britain in the thirteenth century. They are primarily associated with the eastern counties of

England, to the east of a line from the Trent down to the Severn but there were isolated examples in most parts of the country.

There are two types of windmill used for grinding corn, the post mill, which has a central post on which the whole building is turned to face the wind and the tower or smock mill, which has the sails mounted on a rotating top section called a cap. Tower mills were built of brick or stone, smock mills and post mills were built of wood. These windmills provided about sixty horsepower in a stiff breeze. The tower or smock type of windmill was also used for driving wood cutting circular saws and other light industrial tasks.

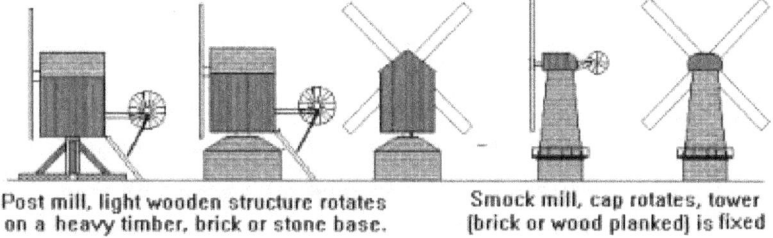

Post mill, light wooden structure rotates on a heavy timber, brick or stone base.

Smock mill, cap rotates, tower (brick or wood planked) is fixed

Figure 6-1: Windmills

Windmills of lighter construction, typically a light metal framework tower topped with a multi-bladed propeller, were widely used for pumping water, either draining flooded land or lifting water from wells.

Water wheel: The history of water wheels dated back to the Roman period when used to grind grains and lift irrigation water, but fell from use when they left until about the eighth century. The number of water wheels then steadily increased until by about the thirteenth century most villages had at least one water powered mill for grinding corn. Later water wheels drove the bellows in the iron works and the spinning machinery in the textile mills. Water wheels are more reliable than wind power although droughts and dams could affect their use.

The British Engineer John Smeaton (1724-1792) did a lot of work on water wheels; he introduced metal gears and axles to water wheels in the mid to last half of the 18th century, also conducted a scientific investigation into the design of water wheels which lead to significant efficiency increases and also proved that the over-shot wheel, driven by the weight of the water, was more than twice as efficient as the undershot type then in common use, which is driven only by the flow of the water. The first all-metal water wheels were built in the 1830's and they gradually displaced the wooden types as these fell due for renewal.

Water mills have remained in use, as they did a perfectly good job and required little ongoing investment there was little need to change.

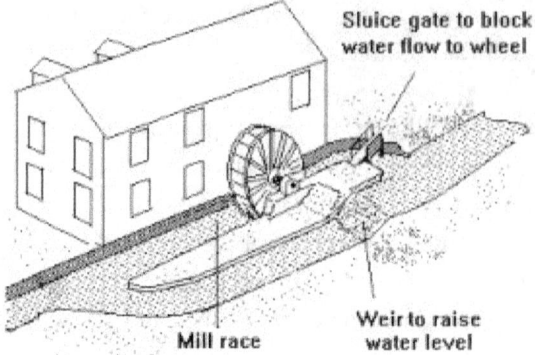

Figure 6-2: Water mill

Water turbine: Water turbines date from the early nineteenth century when a French engineer called Benoit Fourneyron built a horizontal high speed water wheel. Fourneyron went on to build over a thousand of this turbine installations all over the world and Fourneyron turbines are still used today where there is a low pressure flow of water such as a river. In 1889, an American engineer called Lester Allen Pelton (1829-1890) patented an improved design known today as the Pelton Wheel. This is the turbine used for high pressure water turbines which are used where a high head of water is available such as at a waterfall or a purpose built dam.

Pelton turbines are employed in hydro electric power stations. Water power used to drive simple mechanical devices continued in use into the twentieth century but as steam engines were improved they became increasingly popular and steam remained the major source of power into the first half of the 20th century.

Steam engines were developed partly because of the shortage of water supplies to drive water wheels. This was most notable in Cornwall where there is little surface water, and it is no surprise that so many of the pioneers of steam power came from that part of the country.

Horse engine development

Horses became literally harnessed to the ploughs and wagons when the horse collar was widely adopted for use in Britain after about 1700. Later in 1700s, the horse powered engines were developed. A horse engine (also called a horse power or horse-power) is a now-largely obsolete machine for using draft horses to power farm implements and other industrial process machinery. Examples of machines that were powered with a horse engine include the threshing machine, the corn sheller, pumps and machines for fir fighting.

The most earliest of all the horse engines developed was the 'gin' (a shortened form of 'engine') or 'whim', a capstan fitted with a long arm to which the horse was harnessed. The drum was placed well away from the shaft mouth in a separate frame and carried a double ended rope, which was wrapped several times around the drum. The horse walked round and round and the capstan could either wind up a length of rope to pull trucks along a plate-way or, via gears, drive a shaft to power a small mill or machinery such as a corn threshing machine (invented by Andrew Meikle in 1784).

Figure 6-3: Horse gin, 1780s

Where these horse powered gins were used at mines and the like they were usually in the open, on farms, where they powered fixed threshing or milling equipment, they might be enclosed in distinctive circular buildings. By the 1990's there were very few of these buildings left in existence. There is a very nice model of a portable gin on the Pendon Museum 'Vale Scene', it is hooked up via a rigid shaft to an elevator being used to build a hay stack on one of the farms. The example shown in Figure 7- below is a type that was mounted inside a building; the horse would be attached to the harness on the left. This example can be seen at the Manchester Science Museum identical to one used at a mine in the early 19th century.

Figure 6-4: Horse engine

The use of horse engine in fire fighting became popular in the later part of 19th century. Until the mid-19th century most fire engines were manoeuvred by men, but the introduction of horse-drawn fire engines considerably improved the response time to incidents.

Figure 6-5: Hibernia steam fire engine and horses, 1859

Development of horse treadmills

A common design for the horse engine was a large treadmill on which one or more horses walked. At the start of the 19th century, most horse power usage were still exploited on stationary machinery and fitted with simple low speed gearing systems. But as the century progressed, many forms of gearing were developed to increase the speed of machinery. By the 1830's, in America, both portable sweeps and railway treadmills as required had evolved to power the evolving popular groundhog threshing machines.

In 1830, Hiram A. Pitts and John Pitts of Winthrop, Maine, patented an improvement on a railway or tread power, which consisted in the substitution of hard maple rollers under the movable platform, connected by an endless chain, for the old-fashioned leather belt (Figure 6-6). Hiram and his brother, John began the manufacture of these improved powers on a small scale in their native town, and introduced them in the state of Maine and to some extent in other New England states.

Figure 6-6: Horse powered tread mill used in running threshing machines

They became popular for giving power to the "ground-hog" thresher, as the open-cylinder machine was called there. While operating these machines, Hiram Pitts conceived the idea of combining the old "ground-hog" and the common fanning-mill in a portable form. The portable horse-powered treadmill invented was coupled with a thresher, or "separator. In 1834 Hiram completed a machine on this plan which operated successfully. E. Briggs, of Ft Covington, NY was also reported to have invented and patented a horse treadmill on July 12, 1834, for use with his threshing machines.

The primitive chain consisting of U-bolts under each tread with loose connecting links were replaced by the invention of M. Davenport, of Phillips, ME, who patented a wooden cog belt treadmill on Oct 10, 1835.

The horse engines, the treadmills and steam engines were in turn gradually displaced after 1900 with the invention of the petrol powered tractor. However, many small farmers could not afford the cost of a tractor and continued to use horse engines well into the 1940s.

Small animal power

With the inception of the industrial age came the need to find power sources for all sorts of machinery. Steam engines were used for large machines but smaller ones like butter churns needed a more compact form of power. Eventually internal combustion engines and electricity met this need but in the interim, small animals like goats, sheep and of course, dogs, were used. The heyday for this form of animal power was the 1800's.

The treadmill require some sort of platform to provide rotating or reciprocating energy from a dog walking to a form that could be used by the machine as modified by the gearing. The devices used for animal power, especially for dogs, kept getting more and more sophisticated culminating in many patent applications in the 1870's and 1880's. Some of these inventions involved improved footing for the animal and designs to keep the animal in the centre of the track.

Figure 6-7: Treadmill for small animal @ Hadley Farm Museum

Since the treadmill derives their power from the weight of the animal, rather than the draft, it proved particularly useful for smaller animals such as dogs, sheep and goats. For heavy work, horses were hitched so both weight and draft contributed to the power output of the treadmill. Where only the animals' weight was used, the amount of power could be controlled by adjusting the angle of incline of the track.

Figure 6-8: Dog powered butter maker (background is the water pump)

These small treadmills provided both rotary and reciprocating power to operate light machines like butter churns, grind stones, fanning mills, corn shellers, and later, cream separators. They generally use two India rubber or leather belts rather than iron links to form the chain of wood treads. The few surviving dog treadmills have become popular attractions at engine and farm shows across the country. Early treadmills were made almost entirely of wood.

By 1800, steam powered engines had began to replace horses on larger farms. It is perhaps worth noting that in the mid 1980's about half the world was still solely dependent upon human or animal muscle power. Even in Britain horse powered systems remained in use in to the middle of the 20th century, mainly on farms and at smaller mines and quarries.

Steam powered engine development

Historic records reported a Catholic priest named Father Ferdinand Verbiest to have built a steam powered vehicle for a one time Chinese Emperor, Chien Lung at about 1678. However, there was no information about the practical construction or development of this vehicle, only the event. The first practical documentary evidence about steam engine in the primitive era was done in the late 1600s by a French inventor named Denis Papin (1647-1712) who produced the first piston engine; precisely in 1690.

His engines were too inefficient to do any practical work but the idea was later taken up by other notable English inventor; Thomas Savery (1650-1715). Savery patented a primitive steam engine for pumping water from mines precisely in 1698, called '*the miners friend*' it was a simple design and was rather prone to explosion. The engine worked on air pressure, in which steam was blown into the cylinder which was then cooled. This condensed the steam causing a partial vacuum to be created and the air pressure pushed the piston into the cylinder.

Timeline of agricultural mechanization

By 1705, an improved version of this engine known as *atmospheric steam engine* was invented by a Devon-born English blacksmith and inventor, Thomas Newcomen (1663-1729), when working with another Englishman named John Cawley (?-1725). These two inventors partnered with Savery for the application of their invention in pumping water. Newcomen subsequently went into partnership with Savery and their firm introduced the first ideal practical steam engines used to pump water from mines in about 1712. Newcomen's pistons had to be up to two inches smaller than the cylinder and leather washers were used to form a seal. This engine (Figure 6-1) was installed at Dudley Colliery (West Midlands) to trigger changes in the Cornish mining industry.

Figure 6-9: Newcomen atmospheric engine

Newcomen engine was used throughout Britain and Europe, by the time of Newcomen's death in 1729; his engines had helped in draining mines in Hungary, Sweden, France, Germany etc. The Newcomen engine introduced a radically new method of working and also created the necessity for skilled workers who later became known as *engineers*. The Newcomen engine was not effectively efficient but the same basic engine remained in use well into the nineteenth century.

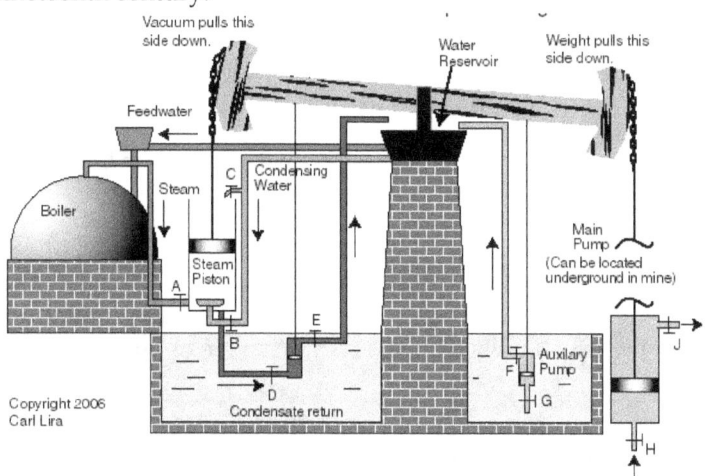

Figure 6-10: Schematic of Newcomen atmospheric engine

Between 1753 and 1755, Josiah Hornblower erected the first beam engine in North America in New Jersey. However, Scot James Watt (1736-1819) in 1765 developed the first pressurized steam engine which proved to be much more efficient and compact than the Newcomen engine. He was famous for inventing an improved version of the steam engine in 1769 and manufactured it till 1819; the year he died. His improved steam engine had a bored cylinder with a close fitting piston and there was a separate condenser connected to the cylinder, allowing the cylinder to remain warm when the steam was condensed. In the mid 1770's James Watt and Matthew Boulton (1728-1809) set up in partnership at Boulton's Soho works near Birmingham and started the production of steam engines.

In 1775 John Wilkinson, a Staffordshire iron maker, developed a practical boring machine capable of boring a near perfect circular hole in solid metal. This was a major advancement when applied to steam engine cylinders. The resulting cylinder was much stronger than the earlier designs based on riveted plates but more importantly, as it was accurately drilled it did not leak badly.

In 1780 Watt built the first steam powered flour mill, obtained a patent for steam engine that provided means of changing the motion of the piston to rotation for driving machinery in 1781 and in 1782 he invented the double acting cylinder, in which steam is admitted at alternate ends to drive the piston in both directions.

In 1788 Watt invented the flyball or centrifugal governor which uses two weighted arms mounted on a rotating shaft connected to the engine. As the speed of the engine and hence the rotation of the shaft increases the arms tend to fly outwards, this motion is then linked to the steam valve to bring the speed back down again. This meant that engine speed would be maintained with changes in temperature in the boiler and with varying loads, amongst the first examples of feed-back control systems this simple mechanism remains in use today. The example to the right in Figure 6-11 below is typical on a machine at the Manchester Museum of Science and Industry.

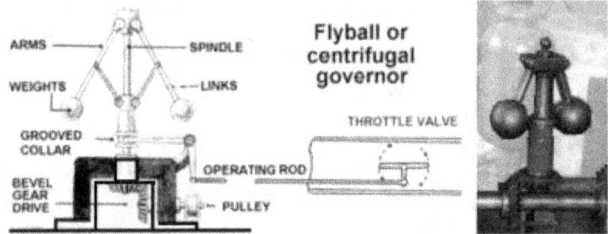

Figure 6-11: Fly ball governor principle of operation

In 1791, British patent on gas engine using coal gas was awarded to John Barber of England, and in 1799, coal gas engine patent that compressed a mixture of gas and air before ignition was awarded a Frenchman, Philippe Lebon (1767-1804). In the early

1800's, Samuel Brown, L. W. Wright, and William Barnett made similar advances in coal gas engine designs and later, its application was extended to moving vehicles. A Cornish mining engineer called Trevithick developed steam engines which used much higher pressures than the Newcomen/Watt types, originally intending to build engines for pumping water out of the Cornish tin mines. He went on to build a steam powered road carriage (in 1801) and later in 1812, a steam locomotive.

The application of steam engine was limited because of the enormous weight that was required in the machine. The introduction of high-pressure boilers in the 1850's did much to lighten engine weights. George Frick started a business to build steam engines in 1853 which remained in operation into the 1940's.

Steam locomotive development

The history of the automobile reflects an evolution that took place worldwide. It is estimated that over 100,000 patents created the modern automobile. Starting with the first theoretical plans for a motor vehicle that had been drawn up by both Leonardo da Vinci and Isaac Newton, the first ever steam engine mounted on a vehicle to move under its own power for which there is a record was designed by a French military engineer, Nicholas Joseph Cugnot (1725-1804) and constructed by a mechanic named M. Brezin in 1769 under Cugnot's instructions at the Paris Arsenal. Cugnot presented his invention named "fardier a vapeur" as an artillery piece in 1769 (AGEON, 1989).

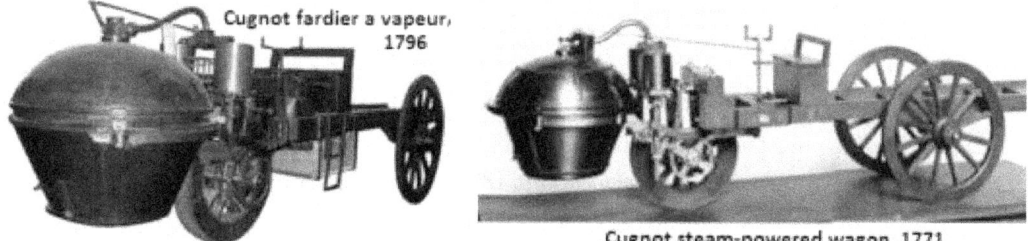

Figure 6-12: Cugnot steam engine

Cugnot's steam engine powered a three-wheeled road wagon built and demonstrated in France in 1769 thereby becoming the first inventor to convert the alternate motion of a steam piston into rotary motion. It was used by the French Army to haul artillery at a whopping speed of 2 ½ mph on three wheels! The vehicle had to stop every ten to fifteen minutes to build up steam power. The steam engine and boiler were separate from the rest of the vehicle and placed in the front

A second unit of his engine was built in 1770 which weighed 8000 pounds and had a top speed on 2 miles per hour, carried four passengers, and on the cobble stone streets of Paris this was probably as fast as anyone wanted it to go. Cugnot also built a second wagon, dating 1771, which was able to transport a 5-ton load at a speed of 3.5 km/h. This was considered a very heavy vehicle, hard to manoeuvre with two rear wheels and one in the front, supporting the steam boiler. The wagon has many constraints which led Cugnot to drive one of the units into a stone wall (Figure 6-13), making Cugnot the first person to get into a moving vehicle accident! This was the beginning of bad luck for the inventor. One of his patrons died and the other was exiled, coupled with the political environment in France which led to the French government ending this project and the money for Cugnot's road vehicle experiments terminated.

Figure 6-13: Autocrash of Cugnot steam-powered wagon, 1771

Following the demonstration of the three-wheeled steam road wagon, Richard Lovell Edgeworth patented a steam engine that traveled upon an endless railway system in 1770. On this machine, flat bearers were attached to the wheels of the engine to support its weight on soft ground. The method was not considered successful until the Boydell patent in 1846.

In 1803, practical steam (vertical boiler) vehicles (carriage) were designed and manufactured by Englishman Richard Trevithick (1771-1833), and his first locomotive in 1812. His No 14 engine was built by Hazeldeine and Co. Bridgnorth, at about 1804, and illustrated after being rescued later at about 1885 (as shown in Figure 6-5 from Scientific American Supplement, Vol.X, No. 470, Jan 3rd, 1885).

Figure 6-14: Richard Trevithick's engine, 1803

In 1807, an American called Robert Fulton (1765-1815) built a successful steam powered ship, this was not the first steam ship but it was the first to be a commercial success. The noise and vibration when the steam engines were used was reportedly most annoying. Trevithick's invention of the locomotive in 1812 was considered the first tramway locomotive ever produced, however, it was originally a road locomotive, designed for a road and not for a railroad.

Figure 6-15: Richard Trevithick's locomotive, 1812

Trevithick's accomplishments were many and the inventor did not fully receive the credit he was due during his lifetime.

The early engines worked by principle of condensing steam in a vacuum, whereas later types (such as steam locomotives) used the power of expanding steam. Between 1803 and 1829, several people in England and France worked on the development of steam powered locomotives with George Stephenson (1781-1848) and Robert Stirling being the most successful.

Robert Stephenson (1803-1859), with his works at Newcastle, was probably the most influential locomotive engineer of the time and his father George was one of the strongest advocates of steam-hauled railways such that he was appointed as the engineer for the Stockton & Darlington Railway of 1821, built primarily to carry coal this was the first successful 'public' railway to use steam engines.

These engines were built as small road locomotives and were operated by one man if the engine weighed less than 5 tons. They were used for general road haulage and in particular by the timber traders. The photo below was scanned from a book published in the 1930s; the photo was taken at the Centenary exhibition on the line in 1929 and shows a replica engine built for the occasion which proved capable of speeds of up to 12 miles per hour when pulling a full load. The telegraph pole with its bank of insulators would not have been present when the original engine was in service.

Figure 6-16: No. 1 replica of the Stockton and Darlington line locomotive

Steam railway locomotives remained ungainly and unreliable machines mainly developed for use on short runs in collieries but progress was rapid. A horse could pull typically four coal wagons at about three miles per hour (5kph), by 1830 steam locomotives were pulling upwards of twenty wagons at anything up to twenty miles per hour (32kph). In 1830 railway engines developed from the successful Rocket began operating on the Liverpool and Manchester Railway and soon proved their worth.

The Stirling gas engine was patented in 1816 by the Scottish clergyman Robert Stirling (1790-1878) and was used as a small power source in many industries during the 19th and early 20th centuries. The Stirling engine is simple in design, a sealed series of chambers contain a quantity of 'working gas'. This starts in a cool chamber where it has been compressed. The gas is then passed to a 'hot chamber' (heated by an external flame) where it expands and drives a piston that delivers the work. The gas is then passed into a radiator to cool down and is then compressed and returned to the original cool chamber. The expansion of the gas at high temperature delivers more work than is required to compress the same amount of gas at low temperature. The heat for the expansion chamber is provided by an external burner supplied with any one of a range of fuels and the exhaust generated has very low free carbon and toxic gas levels.

The Stirling engine runs smoothly but the need for a big radiator makes it unsuitable for motor vehicles. The growing need for motor vehicle engines with relatively clean exhausts has revived interest in the Stirling engine, and prototypes have been built offering up to 500 horse power and with efficiencies of 30 to 45 percent (modern motor car engines have efficiencies in the range of 20 to 25 percent).

In 1845, William McNaught (1813-1881), an American, invented the 'compound engine' or 'double expansion' engine in which the steam is first used to drive a high pressure cylinder, then re-used to drive a second low pressure cylinder. It was not long before a third cylinder was added, producing the 'triple expansion engine'. These were not initially very successful but when forced draft ventilation of boiler fires was introduced, greatly increasing the pressure of steam produced, they became the engine of choice for steam ships. Some years later various railway companies built locomotives with two-stage 'compound' engines, one example being the Midland Railway 4-4-0 as offered by Graham Farish. This locomotive has two cylinders mounted on the outside as normal but with a third (high pressure) cylinder hidden underneath and connected to a cranked axle on the leading pair of driving wheels. In 1859 William Rankine (1820-1872) produced the first comprehensive manual for steam engine design.

Steam engine applications

The steam engines were adapted and evolved for agricultural use. Earliest stationary steam power used for agricultural purposes was made by Boulton and Watt, Ltd., Birmingham, England. Stationary steam engines, used for agricultural and other purposes, were manufactured by James P. Allaire (1786-1858), of New York, NY in 1832 while Cable ploughing machine (powered by stationary steam boiler on wheels, moved by horses) was invented and patented in USA by E. C. Bellinger of South Carolina in 1833. These portable steam engines were typically so large and expensive that a threshing crew of from 8–18 men hauled the machines to separate farms where they threshed until done and then went on to the next farm.

The first engines used to power farm implements in the early 19th century were coal or wood fueled portable steam engines on wheels that could be used to drive mechanical threshers by way of a long wide flexible leather belts.

Steam traction engines

Many of the first steam traction engines of the 1800's were mounted on wheels, transported to the field by horses and standing still would haul ploughs and other implements by cable. In 1810, one Major Pratt patented a steam haulage system in which the plough was dragged on the end of a rope. A steam engine was stationed at one end of a field and an anchor cart with a horizontal pulley was set at the other end. A winding drum beneath the engine's boiler turned an endless rope around the anchor pulley, alternately playing it out and winding it back onto the drum. The plough, attached to the rope, moved between engine and anchor.

The most popular of the steam traction engines was the Garrett 4CD. The steam traction engine was heavy and sank in fields that were even slightly soft, so it was restricted to highway transport and powering of threshing machines at harvest. Eventually, the steam traction engine was equipped with a cable drum that could winch a plough or cultivator across the field, and this was the nearest it ever came to the land.

Figure 6-17: Steam traction engine

British soil scientists welcomed these innovations for they prevented soil compression associated with horse cultivation. The aim of farmers though was to create a machine capable of pulling implements or cultivating the soil directly, turning the soil more efficiently than a traditional plough. To encourage such a development, in 1854 the Royal Agricultural Society of England offered a 500 pound prize for the most efficient and economic steam cultivator.

Figure 6-18: Russell steam powered tractor

Inventors competed for four years until an innovation was found worthy and the prize was finally awarded. John Fowler was the winner, presenting a machine that could draw an agricultural implement, like a plough, back and forth across the field using a windlass (Figure 6-19).

Figure 6-19: Fowler´s traction engine pulling a plough (Source: Steam Plough Club)

It was not until the 1854 that the steam engine was first used in ploughing in Europe and exhibited by John Fowler (1826-1864) in England. John Fowler was the first to show that cable cultivation could be profitable. His apparatus used three systems; single engine, double engine, and roundabout. His idea of using a stationary steam engine to wind in a rope and draw a mole plough won the prize of 500 pounds in 1858. The steam plough, although being able to plough ten times the area that horses could plough in a day, was disadvantaged by heavy weight, cumbersome, costly, frequent boiler explosions caused by low water and other factors, and had only a limited impact on farming in either Europe or the United States.

Another inventor, Thomas Rickett, developed a rotary steam cultivator, which was an enormous 10 ton machine designed to break and aerate the soil, that became popular at the time. Between 1858 and 1860s, portable self-moving steam traction engines were developed by Englishman Thomas Aveling and in 1877 self-propelled traction engine was produced by J.I. Case Threshing Machine Co. Case became the most important steam traction engine ever produces in the world at that time.

Steam traction engine development

The steam traction engine evolved from the steam locomotive. The first steam powered traction engines were developed and adopted and was introduced to farmers for agricultural use on large farms in 1868. In 1869, J.I. Case designed an 8hp steam engine used in threshing operations, power drainage pumps and sawmills belt operation. The steam engine was pulled by horses to the site and a belt attached to the large wheel on the left of the machine. Thus the horse remained the mains source of power until the early twentieth century.

Figure 6-20: J.I. Case steam engine, 1869

In 1887, steam-powered traction engine (later called tractor) was patented and manufactured by Daniel Best (1838-1923), who also invented a combined steam-driven harvester and thresher. John Froelich (a custom thresher operating in South Dakota) experienced problems with his J. I. Case steam engine which prompted him to obtain a

Van Duzen vertical single cylinder gas engine with a bore and stroke of a whopping 14". It was considered that had the cylinder been placed horizontally, the end motion caused by the giant piston would have rendered the unit inoperable. Ignition was by a wet cell battery and contact points. Froelich mounted the engine on a Robinson steam engine chassis. He devised a reversing gear so that it could back up.

Figure 6-21: Froelich steam-powered traction engine, 1892 (IMJ archives)

Froelich used his engine during the 1892 threshing season. He formed the Waterloo (Iowa) Gasoline Traction Engine Co. Between 1893 and 1896 he produced four machines. Two were sold but subsequently returned because they did not meet the buyer's expectations. The Waterloo Co. made a corporate decision to concentrate on stationary engines, and they did not re-enter the tractor business until 1913. Froelich left the company which was reorganized as the Waterloo Gas Engine Co.

American Gasoline Engines considered Benjamin Van Duzen a pioneer in the development of the internal combustion engine since 1872. Operating as the Van Duzen Gas & Gasoline Engine Co., in Cleveland, Ohio, Van Duzen received more than 15 patents for his designs. In 1890, Van Duzen features a unique 6 HP valve motion on both the intake and exhaust side of the engine.

Figure 6-22: Van Duzen engine, 1890 (Williams, 2012)

In 1894, the company produced an engine for their own tractor, but with different characteristics to the engine supplied to Froelich. The engine is a 10-inch stroke igniter and Wizard-type friction drive magneto with a hit-and-miss governing mechanism. It was started by a shot gun cartridge in a breech (similar to later Marshall and Field Marshall tractors) and then relied upon a platinum tube to retain the heat for ongoing ignition.

Figure 6-23: Van Duzen design, circa 1914 (IMJ archives)

There are speculations that he may have worked for Hart-Parr after 1901, but there has not been documentary evidence on this claim.

In 1898, George B. Geiser (1870-1938) of the Geiser Manufacturing Co. built a steam engine known as Geiser Peerless steam engine. The engine enjoyed its largest patronage between 1885 and 1914.

Figure 6-24: Geiser steam engine

Shown in Figure 6-25 below is the rear view of a Geiser steam engine in 1908 pulling a twelve bottom plough on highland farm Fullerton, in North Dakota. Beside the tractor is a water wagon with hoses running to tractor; three unidentified men are visible around the tractor. In background is the Highland farm.

Figure 6-25: Peerless plough on highland farm Fullerton, N. Dar

The gas traction engine development

J.I. Case's first pioneering efforts at producing a gas traction engine dated to 1894, or maybe earlier, when William Paterson of Stockton, California, came to Racine to make an experimental engine for Case. Case's historical records in the gas tractor field, claimed 1892 as the date for Paterson's gas traction engine: the patent date suggested 1894. The early machine ran, but not well enough to be produced.

These vehicles ultimately became so useful and resourceful that farmers took to calling them simply GPs, for general purpose. But they were not always so highly regarded. Early versions, powered by bulky steam engines, lumbering along on steel wheels, some weighing nearly 20 tons, often mired in wet and muddy fields—practically worthless. Charles W. Hart and Charles H. Parr began their pioneering work on gas engines in the late 1800s while studying mechanical engineering at the University of Wisconsin at Madison. In 1897, the two men formed the Hart-Parr Gasoline Engine Company of Madison.

In 1900, they moved their operation to Hart's hometown of Charles City, Iowa, where they found financing to make gas traction engines based on their innovative ideas. Their efforts led them to erect the first factory in the United States dedicated to the production of gas traction engines.

Early internal combustion engine development

In 1678, during the reign of Louis XIV, a French monk named Jean de Hautefeuille (1647–1724) invented a method of producing energy by repetitively burning a measured amount of black gunpowder as a fuel within a cylinder thereby proposed an early form of internal combustion engine. It seems unlikely that any such machine was ever constructed by him. He was, however, the first person to propose the use of a piston in a heat engine.

Christian Huygens (1629-1695), a famous Dutch physicist proposed a similar device two years later in 1680 based on de Hautefeuille's suggestion and appears to have constructed some form of its prototype. He built a machine with used gunpowder to push a piston up a cylinder, falling again by its weight and by air pressure as the explosive gasses cooled and contracted. He was never able to solve the problem of repeating explosions.

Other Europeans developed the principle of gunpowder engines throughout the 18th century. But the big breakthrough in the design of the internal combustion engine occurred in 1859, when Edwin C. Drake drilled the world's first commercial oil well at Titusville, Pennsylvania thereby making available petroleum spirit.

Contrary to operation of steam engines operating basically on the principle of an injection of steam created in a boiler, which is remote from the actual engine, and being forced under pressure into a cylinder, the internal combustion engine is one in which the creation of energy occurs by the expansion of heated gases within the engine. Steam engines and the Stirling engines had been classed as 'external combustion engines', in that the fuel is burned on the outside of the engine. The difference between a steamer and an internal combustion engine is that the latter obtains its energy from heated expanding gases, the entire process occurring within the engine, without the necessity of a remote boiler and heat source.

Internal combustion engines, which burn the fuel inside the engine, were actively researched from the mid 17th century. Prior to the availability of petroleum spirit, turpentine was the first liquid to be used as a fuel in an internal combustion engine. This occurred in 1794 when an Englishman Robert Street discovered the volatility of this distilled tree oil and designed an engine to run on vaporized turpentine fuel. He obtained a patent on the device; this was the first internal combustion engine to use a liquid fuel (it pre-dates the invention of coal gas by Murdoch in the late 1790's). In 1831 William Barnett of Brighton patented an engine which was intended to use hydrogen gas as a fuel.

In 1859, commercially successful internal combustion gas (widely known as coal gas or illuminating gas) engine (operating on two cycle, one cylinder) were developed by a Belgian-Frenchman; Jean-Joseph Earnest (Etienne) Lenoir (1822-1900) in Paris, France. Lenoir patented the first practical gas engine in Paris in 1860 and drove a 'car' based on the design from Paris to Joinville in 1862. Lenoir claimed to have run the car on benzene and his drawings show an electric spark ignition. If so, then his vehicle was the first to run on petroleum based fuel, or petrol, or what we call gas, short for gasoline. His one-half (½) horse power engine had a bore of 5 inches and a 24 inch stroke. It was big and heavy and turned 100 rpm. Lenoir had a separate mechanism to compress the gas before combustion.

In 1862, Alphonse Bear de Rochas (1815-1891) figured out how to compress the gas in the same cylinder in which it was to burn. This process of bringing the gas into the cylinder, compressing it, combusting the compressed mixture, then exhausting it is later known as the Otto cycle, or four cycle engine. Earnest Lenoir and Beau de Rochas did a lot of basic works on internal combustion engines. Lenoir died broke in 1900!

Records showed that Siegfried Marcus, of Mecklenburg, built a car in 1868 which was on display at the Vienna Exhibition of 1873. His later car called the *Strassenwagen* had about ¾ horse powers at 500 rpm. It ran on crude wooden wheels with iron rims and incorporating a brake system operated by pressing wooden blocks against the iron rims to stop the car, had a clutch, a differential and a magneto ignition. One of the four cars built by Marcus is in the Vienna Technical Museum and can still be driven under its own power.

Records had equally shown that steam powered cars had been built in America before the Civil War but the earliest ones on record were like miniature locomotives. In 1871, Dr. J. W. Carhart, a professor of physics at Wisconsin State University, and the J. I. Case Company built a working steam car. It was practical enough to inspire the State of Wisconsin to offer a $10, 000 prize to the winner of a 200 mile race in 1878 which was won by Oshkosh car which finished with an average speed of 6 mph. From this time until the end of that century, nearly every community in America had a *'mad scientist'* working on a steam car. Many old news papers tell stories about the trials and failures of these would be inventors.

In 1876, a German engineer named Nicholas August Otto (1832-1891), developed a practical engine to run on coal gas, which is known as *Otto cycle engine* today. Early forms of internal combustion engine invented were patented in 1877 by Otto and another German; Eugen Langen (1833-1895). Otto obtained a U. S. patent in 1876. His coverage was very broad, and he put a high price on the royalty for his patent. This stifled development until the European courts limited his monopoly in 1890. Otto died in 1891 and his U. S. patent expired in 1894. The 'Otto cycle' four-stroke engine has formed the basis of all subsequent internal combustion designs (with the exception of the 'gas turbine' engine).

Figure 6-26: Otto-Langen gas engine, 1867

In 1876, when Otto patented the Otto cycle engine, de Rochas had neglected to do so, and this later became the basis for Daimler and Benz breaking the Otto patent by claiming prior art from de Rochas. In 1885 the German engineer Karl Benz (1844-1929) produced the first practical three wheeler motor car, using a gas engine of the Otto type. Also in 1885 Gottlieb Daimler (1834-1900) patented his engine which could run on petrol or gasoline and which ran at much higher speeds than earlier types, offering greater power.

Figure 6-27: Gottllieb Daimler's workshop, 1885

Figure 6-27 shows the picture of Gottllieb Daimler's workshop in Bad Cannstatt where he built the wooden motorcycle shown. Daimler's son, Paul was reported rode this motorcycle from Cannstatt to Unterturkheim and back on November 10, 1885. Daimler used a hot-tube ignition system to get his engine speed up to 1000 rpm. The previous August, Karl Benz had already driven his light, tubular framed tricycle around the Neckar valley, only 60 miles from where Daimler lived and worked. They never met. History recorded that Frau Berta Benz, Karl's wife, took Karl's car one night in 1888 traveled 62 miles from Mannheim to Pforzheim to see her mother, and made history of making the first long car trip!

In 1889, Daimler built a two-cylinder engine which soon proved its worth in races and was adopted by several motor manufacturers. Also in August 1888, William Steinway, the owner of Steinway & Sons piano factory, talked to Daimler about US manufacturing right and by September had a deal. By 1891 the Daimler Motor Company, owned by Steinway, was producing petrol engines for tramway cars, carriages, quadricycles, fire engines and boats in a plant in Hartford, CT. In 1889, Franz Burger and the Charter Co. developed an engine on wheels that could be moved about for belt work. Other inventors began to experiment in 1890's; the year Ransom E. Olds built his second steam powered car (Figure 6-28). One was sold to a buyer in India (Figure 6-28 left), but the ship it was on was lost at sea. Ransom E. Olds was adjudged the first mass producer of gasoline powered automobiles in the United States.

Figure 6-28: Olds steam automobile, 1890; right: Olds first petrol powered car, 1896

In 1892, Dr. Rudolph Diesel (1858-1913) experimented with an engine fuelled with coal dust. The dust was blown into the combustion chamber by compressed air. But, subsequently were a number of unplanned and somewhat alarming explosions; he concluded that the challenges of perfecting the coal dust engines were proving to be too dangerous in relation to any possible benefits. Thereafter, Diesel directed his attentions to the possibility of using crude oil as an alternative fuel.

In 1894 he successfully injected crude oil under pressure into a combustion chamber, in which the air had been compressed. The resultant pressure detonated the fuel and thus created compression ignition. The diesel engine had arrived! The Diesel engine did away with the need for a spark plug and its associated unreliable electrical equipment and allowed heavier, less refined, fuels to be used. At about the time of the First World War the 'semi-diesel' engine appeared. This is a diesel engine which develops a 'hot spot' in the cylinder (some were equipped with an electrically heated bulb) and this allows lower grade fuels to be used. Semi-diesels often needed pre-heating with a blow-lamp to get them started but the big advantage was that they would run on almost any fuel, including waste engine oil diluted with paraffin. Semi diesel engines were used on agricultural tractors into the 1950s.

Charles W. Hart and Charles H. Parr met as mechanical engineering students in 1892 at the University of Wisconsin in Madison, Wisconsin. Their mutual interest in the development of the internal combustion engine led them to a joint, extra credit project to produce an internal combustion engine. In fact, during this period, they produced five engines and graduated with honors. Their first engine used a push-rod operated exhaust valve and a spring loaded intake valve. Speed was controlled by holding the exhaust valve open when the engine reached its upper limit. The engine would then miss-fire. This was called the "hit or miss" system.

By 1901 there were twelve examples of IC engines mounted on wheels, some of which could be called tractors. Hart and Parr introduced the valve-in-head design with overhead cam in their 1896 stationary engine. During the 1901-04 periods they used both push-rod and overhead cam operated valves. They referred to the overhead cam operated valve as a rotary valve. After 1905 the overhead cam was used exclusively. This is standard in almost all IC engines manufactured today.

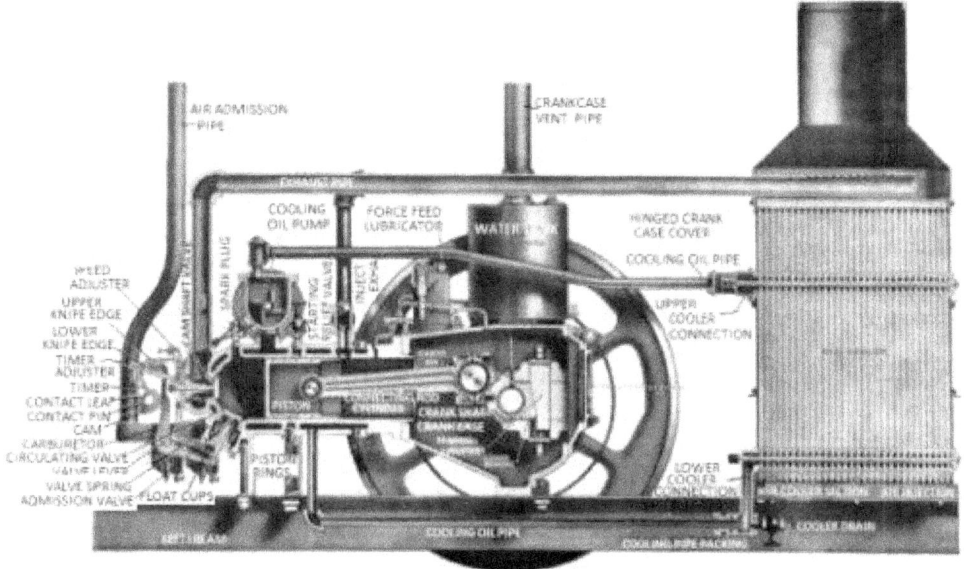

Figure 6-29: Sectional view of Hart-Parr 30-60 engine

The introduction of the overhead cam and valves in the head brought the horsepower up to 30 and 60 on the drawbar and pulley respectively. Designated the 30-60, it became known as the "Old Reliable" and it was produced until 1918.

Figure 6-30: The old reliable 30-60

In 1903, Dan Albone, a bicycle manufacturer in Biggleswade, England, produced a lightweight tractor called the *Ivel*, compared to the early American tractors. The tractor was powered by a two-cylinder horizontally opposed water cooled engine. The Ivel was the first British tractor to be exported in large numbers. In the same year, the oldest tractor manufactures in Australia appeared in market. The sample pictured in Figure 6-31 is Australia's oldest tractor and is owned by Norm McKenzie.

Figure 6-31: Australia's oldest tractor, 1903 (Photo IMJ)

Hart-Parr introduced the heavier drive gear design in 1904, called a plough gear. The standard gear used a bevel gear differential. A spur gear differential was used for the plough gear. This could transmit more torque. The same tractor models could be ordered with either differential. Hart-Parr tractors were known for their rugged construction.

In 1907, the Transit Thresher Co. of Minnesota used its own innovative design of four cylinder engine, with a 6x8 inch bore and stroke, in its Big Four at a time when other manufacturers favoured more simplistic single and twin cylinder engines. The firm was restructured in 1908 and became the Gas Traction Co.

Figure 6-32: Hart-Parr transit thresher's four cylinder engine (IMJ archives)

The engine was eventually upgraded with a tubular cooling radiator replacing the thermal siphon system, and in 1912 came under the control of the Emerson Brantingham Implement Co. of Rockford, Illinois.

Figure 6-33: Big 4 tractor pulling a plough

Marshall and Sons of Gainsborough, England, introduced its Class C two cylinder engine tractor in 1909. Note the flying governor and the massive flywheel. The two cylinder unit produced around 30 hp. Figure 6-34 shows a restored version owned by the Pioneer Village at Swan Hill, Victoria. The large tank above the front axle is a thermal siphon water cooling reservoir. The tractor was provided with a single speed gear.

Figure 6-34: Left: Marshall Class C engine, 1909; Right: Restored version (Photo M Daw)

In 1912 an American by the name of Charles Kettering (1876-1958) invented the electric self-starter for internal combustion engines so drivers no longer had to crank the engine by hand, risking a broken arm if it back-fired. Petrol engines were still temperamental, one big problem was 'knocking'[18] and the oil companies experimented widely to try and eliminate this problem. Also in 1912, Rumely's two cylinder Oil Pull engine was manufactured complete with clutch, lubricator and magneto as shown in Figure 6-35.

---

[18] Knocking occur when little quantity of fuel remain unburned on the combustion stroke, this then explodes under compression during the exhaust stroke. One idea which caught on was that the colour of the fuel could be a factor and several companies added dyes to their petrol. The origin of this theory is unclear; it may have come from experiments in which a catalyst was present in the dye used. In 1921 Thomas Midgley (1889-1944) discovered that adding tetraethyl lead to petrol eliminated the knocking and this was then added to all motor fuels until the 1980's.

Figure 6-35: Rumely's two cylinder Oil Pull engine (left); tractor (right)

The 30-60 Rumely Oil Pull tractor was a true heavyweight with an impressive reputation for reliability. Weighing around 12 tons, it featured 80 inch tall rear wheels mounted on 5.475 inch diameter rear axles. The twin cylinder engine produced 30 drawbar hp and 60 belt hp. In the same year, 1912 the Lanz Landbaumotor was powered by a four cylinder Kamper petrol engine manufactured in Berlin. The attached Lanz rotary cultivator was possibly the first successful type produced commercially. It was chain driven and somewhat remarkably incorporated a hydraulic lift mechanism. Devotees of early tractors will note the similarity of the Lanz Landbaumotor to the American Wallis Bear.

Figure 6-36: Four cylinder Kamper petrol engine (Courtesy Lanz archives, Mannheim)

Gasoline/petrol-powered engines

Charter Gasoline Engine Company of Chicago, Illinois, was credited first in successfully using gasoline as fuel. Charter's creation of a gasoline fueled engine in 1887 with the power unit built on a Rumely steam traction engine frame soon led to early gasoline traction engines before the term "tractor" was coined by others. The earliest internal combustion engines were stationary and could power barn equipment like threshers and other machinery. Some were mounted on wheels so they could be transferred from field to field.

Figure 6-37: First successful Hart-Parr gasoline engine, 1901

Until 1904, Hart-Parr engines were designed to run on gasoline. They originally used a fuel pump system, but because of a lack of gasket and packing materials that would tolerate gasoline without leaking, they substituted a gravity fuel feed system. They had designed their own carburetor. Although Hart and Parr failed to take the trouble to caver many of their inventions with patents, they did apply for coverage on the double carburetor in 1908 which was issued in 1915 when Meinard Rumley, a blacksmith in LaPorte, Indiana had copied much of Hart- Parr technology.

Charter adapted its engine to a Rumley steam-traction-engine chassis, and in 1889 produced six of the machines to become one of the first working gasoline traction engines (Sanders, 1996).

Kerosene powered engines

Gasoline supplies were becoming less reliable by 1904 because of competition from the rapidly developing automobile market. Prices for gasoline were increasing, and there was speculation that it might be rationed. Kerosene was cheaper, more plentiful, and the energy content per gallon was higher because kerosene is denser.

Consequent on above reasons, in 1904 Hart and Parr invented a double carburetor that metered kerosene through one side and water through the other side. An engine running on plain kerosene had a very serious combustion knock problem. Hart and Parr found they could ameliorate this problem by adding water to the combustion air. In practice the engine was started on gasoline. When thoroughly warmed up, the operator changed a valve that switched the fuel from gasoline to kerosene. Then, when the engine began to knock, the operator opened a water valve that provided water injection to the engine. This controlled the knock.

Since the engine speed was controlled by the hit or miss method, water was consumed only when the engine was under load. It was said that water consumption was about equal to fuel consumption. This was the first known use of water injection in an IC engine to control combustion knock. The principle has been used in high-performance engines since then.

Gas turbine or 'jet' engine development

An alternative internal combustion engine is the gas turbine or 'jet' engine. The word turbine was coined by a French engineer Professor Burdin, who was studying water wheels at the Ecole de Saint-Etienne, for a new type of horizontally mounted water wheel. The word was then used by the British engineer Charles Parsons to describe his steam powered engine (1884). He developed the engine for marine use and startled everyone at a Royal Fleet Review in 1897 by outrunning everything the Navy had. The photo below was taken a few years later; it was scanned from a book on engineering published about 1930s.

Figure 6-38: Gas turbine or 'jet' engine

The Parsons engine used steam from a separate boiler to drive the turbine and Parsons believed that an internal combustion turbine were impossible to build. This was proved wrong by another British engineer called Frank Whittle who published a paper on the subject shortly before World War II. In the event it was the Germans who first produced a working gas turbine, they used it to power aircraft for which the very high power to weight ratio is a major advantage. The gas turbine jet engine has replaced piston engines for larger aircraft and most helicopters but it is less suited to more mundane tasks.

Induction (electric) motor development

The electric motor was developed in the early 19th century, Michael Faraday produced a proof of concept model that rotated in 1821, A Hungarian called Jedlik apparently built a vehicle of some kind driven by an electric motor in the later 1820s, the British scientist William Sturgeon built a practical DC motor in the early 1830s and an American Thomas Davenport patented a commercially viable version a few years later. These all relied on battery supplies and were not a commercial success, electrical power was first applied to machines in Vienna in 1873. The limitations of battery design restricted the practical use of electrically powered road vehicles but track based systems such as trams and railways were able to use electricity delivered via the rails or from overhead wires.

The first serious use of electricity for a standard gauge railway locomotive was a battery powered demonstration loco built for the Edinburgh & Glasgow Railway in 1842. After developments in the USA and Germany a short narrow gauge electric railway was operated in the Brighton area from about 1883. By the late 1880's Nikola Tesla's work on alternating current had made three phase power transmission over long distances practical and power stations were being built in cities around the world. Since the late 1920's there has been an increasing trend toward using electricity to power industrial machinery. At the start of the First World War only about a quarter of the industrial processes were electrically powered, by the time of the Second World War this had risen to over two thirds. Electricity is one of the easiest fuels to distribute and it can power small inexpensive motors attached to individual machines, offering many advantages for industry.

Hydraulic power development

A further option for driving machines is to use pressurized water (hydraulics) or compressed air (pneumatics), the main advantage of both of these systems is safety as there is no danger of electrical sparks and no risk from leaking gas or oil. A key figure in the development of hydraulic engines was Sir William Armstrong a lawyer turned inventor living in Newcastle. It was Armstrong who built the first really practical hydraulic machine the hydraulic multiplier or 'jigger'. A jigger is a long cylinder with a piston inside and the piston rod extending through the end of the cylinder. Water is used to drive the piston back and fore and ropes or chains attached to the end of the piston rod or 'ram' then operate the machinery.

Figure 6-39: Hydraulic multiplier 'jigger'

Armstrong's firm built the first hydraulic cranes on a Newcastle quayside in 1846, using water supplied from the local water mains. The example below is on display at the Manchester Science Museum. On the or 'jigger' the pulley system amplified the distance the lifting rope moved for a given movement of the piston, this also meant that the load moved up or down a lot faster than the piston.

The jiggers could be mounted either inside or outside; on free-standing hydraulic cranes they were mounted inside the base of the crane. Where the crane or hoist was mounted on a building they were usually bolted to a wall close by the lift (on the outside of buildings it was typically set vertically into a recess).

6.2 Historic development of agricultural tractors

Introduction

The early history of the steam engine owes a lot to the development of the tractor. As far as we could trace back, the first ever steam engine mounted on a vehicle, dated to 1769, when Joseph Cugnot, the French military engineer presented his "fardier a vapeur" (AGEON, 1989) and the second wagon, dating 1771. This machine remains preserved today at the Conservatoire National des Arts et Métiers, in Paris.

Figure 6-40: Cugnot fardier a vapeur

Cugnot's inventions and those of several others span almost a century until the emergence of Charles Hart and Charles Parr in 1896.

Emergence of the name 'tractor'

Hart-Parr was credited with the coining of the word "tractor" for machines that had previously been called gas traction engines. At the most basic level, a tractor is an engine on wheels with a seat for the operator and a way to attach implements like ploughs, planters and harvesters. A tractor is otherwise a vehicle specifically designed to deliver a high tractive effort (or torque) at slow speeds, for the purposes of hauling a trailer or machinery used in agriculture or construction. Most commonly, the term tractor is used to describe a farm vehicle that provides the power and traction to mechanize agricultural tasks, especially (and originally) tillage, but nowadays they were employed in a great variety of tasks.

The modern tractor had evolved through the development and adaptation of earlier forms of mechanical land powered machines. The development began when farmers recognized the need for more reliable power than draft animals. When tractors first developed, they were huge, heavy, had limited power and cost way too much for all but the biggest farmers to buy.

Steam tractors

Initially tractors were powered by steam and were typically so large that they could only be profitably used on very large farms. Agricultural implements may be towed behind or mounted on the tractor, and the tractor may also provide a source of power if the implement is mechanized. The earliest tractors were the steam-powered ploughing engine (e.g. Geister steam plough tractor. See Figure 6-25). They were used in pairs, placed on either side of a field to haul a plough back and forth between them using a wire cable. Figure 6-41 shows an early steam tractor, towing a living van and a water cart; Note the large black pulley near driver used for powering threshing machines.

Figure 6-41: Ransomes, Sims & Jefferies Ltd 6nhp *Jubilee* of 1908

Where soil conditions and farm size permitted (as in the United States) steam tractors were used to directly haul ploughs, but in the UK and elsewhere ploughing engines were nearly always used for cable-hauled ploughing instead. The tractors were designed to pull implements originally designed and built to be pulled by teams of horses. Some steam-powered agricultural engines remained in use well into the 20th century until reliable internal combustion gasoline and diesel engines were developed (John Deere, 2003).

Figure 6-42: David Roberts tracked steam tractor c.1908

Development of gasoline fuelled tractor

In 1892, Van Duzen Gas and Gasoline Engine Co., OH built the first practical 20hp gasoline/petrol-powered traction engine in Clayton County, Iowa, in USA for the inventor John Froelich; the forerunner of the Waterloo Boy traction engine later manufactured by John Deere & Co. (John Deere, 2003, Xulon, 2002). Features of this engine include; single cylinder, an included clutch, forward and reverse gears. John Froelich, a custom thresherman from Iowa, decided to try gasoline power for threshing. He mounted a Van Duzen gasoline engine on a Robinson chassis and rigged his own gearing for propulsion. Froelich used the machine successfully to power a threshing machine by belt during the fifty-two day harvest season of 1892 in South Dakota.

Figure 6-43: 4 HP Van Duzen engine, 1914

After receiving a patent for the engine, Froelich started up the Waterloo Gasoline Engine Company, investing all of his assets, which by 1895, was all lost and his business resigned to become a failure (Miller 2003). Froelich's machine fathered a long line of stationary gasoline engines and, eventually, the famous John Deere two cylinder tractors.

Figure 6-44: Froelich tractors, 1892 (Janssen, 1996)

Lighter tractors became possible with the manufacture of reliable internal combustion engines around 1900 and in 1902 a tractor powered by internal combustion engine that ran on gasoline was introduced. The first truly mobile tractors were powered by these engines and were light enough (smaller and lighter than its steam-driven predecessors) to go into the fields. They could pull ploughs and operate threshing machines, and ran all day on a single tank of fuel.

Hart-Parr contributions

After several years of successful innovations using steam, Charles W. Hart (1872-1937), and Charles H. Parr (1868-1941) in 1902, built their first factory in USA to manufacture gasoline traction engines driven by an internal combustion engine. During the winter of 1901-02, Hart and Parr built their first gasoline traction engine, which they named Hart-Parr no. 1, Serial No. 1205.

Figure 6-45: The oldest operating Hart-Parr No. 1 tractor

It was built with a 2-cylinder horizontal engine, 9-inch bore by13-inch stroke and was rated at 17-30 drawbar and belt horsepower respectively. It was sold to a farmer in nearby Clear Lake, Iowa. The Hart-Parr No. 1 tractor was used and finally scrapped shortly after World War I.

Figure 6-46: Hart-Parr Model-2 tractor

Hart-Parr no. 2 (also built with a 17-30 rating) was built in 1902. A few number of model-2 tractors still exists in private hands today. For instance, the tractor with Serial No. 1341, owned by William Peterson of Lowell, Indiana was reported to still be in good condition.

Those units were followed in 1903 by production of 14 gasoline traction engines with the 17-30 hp rating (including Hart-Parr no. 3), and 24 units of gasoline traction engines with a larger 10-by-15-inch bore and stroke rated at 22-40 hp.

Figure 6-47: Hart-Parr 18-36 Model tractor

The Hart-Parr 22-45 and 30-60 tractors featured a 2-cylinder horizontal engine with a 10-by-15-inch bore and stroke. These tractors weighed 19, 000 pounds and were used primarily to plough the prairies in the northern Great Plains and power threshing machines.

Figure 6-48: Hart-Parr 22-45 Model tractor

The Hart-Parr 15-30 and 20-40 models were Charles Hart's first attempt at designing a smaller tractor for Midwest farms. Although they weighed 15,700 pounds with a 2-cylinder vertical engine, they were 17 percent lighter than the 22-45 and 30-60. Only three of the 22-40 models were built with a heavy-duty drive train, which enabled them to pull a plough. The 20-40 is shown here in Figure 6-49 below.

Figure 6-49: Hart-Parr 20-40 Model tractor

Charles Hart's second attempt at a smaller tractor was the Hart-Parr 12-27 (re-rated to an 18-35), powered by a 1-cylinder vertical engine and weighing in at 11,000 pounds. Steel castings were used to reduce the number of parts required to build the tractor and thus the cost of the tractor.

Figure 6-50: Hart-Parr 18-35 Model tractor

Charles Hart's Little Devil tractor was a masterpiece of simplicity in the design of a small tractor. Initially, it weighed only 5,000 pounds and could perform many functions, including pulling a cultivator in a corn field. However, the Little Devil (Figure 6-38) was not a success in the field and production was stopped after building 725 of the tractor units. All of the units were sold and none were returned.

Figure 6-51: Hart-Parr little devil Model tractor

As a result, the Hart-Parr Co. formed in 1905 has been recognized by the American Society of Agricultural and Biological Engineers and the American Society of Mechanical Engineers as builder of the first commercially successful tractor with an internal combustion engine.

Hart-Parr's major accomplishments in the development of a successful farm tractor included the oil cooled engine, the valve in head principle with overhead cam, the magneto ignition system, the plough gear, the vaporizing carburetor with water injection, and forced fed lubrication. All of the tractors produced around 1910 were similar in appearance to the Hart-Parr models.

Hart-Parr Co. specialized in manufacture of gasoline traction engine until Henry Ford got into the picture in 1907. Sir Henry Ford produced his first experimental gasoline powered tractor in 1907, under the direction of chief engineer Joseph Galamb. In 1908, Ford introduced the legendary Model T automobile. It was referred to as an "automobile plough" and the name tractor was not used.

Figure 6-52: Henry Ford Converted Model T

In 1909, Meinrad Rumley introduced and tested the 33.5 bhp Rumely Oil Pull tractor designed by John A. Secor (1847-??), built by M. Rumely Co., LaPorte, IN, at the Winnipeg Trials,. Hart-Parr also claimed to have built the first Oil Pull, perhaps the engine used by Rumely.

Figure 6-53: Rumely's Oil Pull tractor

In 1910, Rumely Co. built the 33 hp Kerosene Annie internal combustion tractor, with an engine that started with gasoline and ran on kerosene. It had a throttle governor and used water injection, it was oil-cooled, and ran at a single, governed speed (in contrast with the earlier hit-and-miss engines). In the same year, Twin city tractors were manufactured by Minneapolis Steel & Machinery Co., Minneapolis, MN

Figure 6-54: Rumely's Oil Pull tractors (1915 model M, 8,750 pounds,)

After 1910, gasoline powered tractors were used extensively in farming. In 1911, Electric starter was introduced by Charles F. Kettering (1876-1958) and was first used in automobiles, later in tractors. Between 1911 and 1912, M. Rumely Co., La Porte, IN, acquired Advance Thresher Co., Battle Creek, MI (established in 1885) and the Gaar-Scott & Co., Richmond, IN (established in 1836), forming the Advance-Rumely Co. in 1915. The Allis-Chalmers Co. acquired most of the assets of Advance-Rumely Co. later in 1931.

In 1912, Four-wheel drive (4WD) tractors entered the market: first, the 28 hp Olmstead (Great Falls, MT), with chain drive, and then the Nelson (Boston, MA), with chain drive and four-wheel steering, available in three models, 15-24 hp, 20-28 hp, and 35-50 hp. Same year, tractors were built using anti-friction bearings; one was patented in 1912 by Clarence Alvin Henneuse (1879-1939), while working with Best Tractor Co. Also, same year C. L. Best Gas Traction Co. commenced the production of 70 hp track-laying tractors.

Figure 6-55: Hackney Auto-plough tractor, 1912

The 8-16 hp Mogul tractors and the 10-20 hp Titan tractors, both of which used kerosene fuel were built in 1914 by International Harvester Co. (IHC), followed by a series of other tractors under those names including the IHC 15-30 (also called McCormick-Deering).

Figure 6-56: Titan 10-20 hp, 1918

By 1915, the first crawler-type tractor was developed under the na Cletrac and was first sold in 1916 by Cleveland Motor Plough. Co., Same year, Ford Tractor Co., Minneapolis, MN, produced a Model B Ford gasoline fueled tractor. Its performance led to the enactment of the *Nebraska Tractor Test Law*. Between 1915 and 1916, tractor catalog listed 62 different manufacturers giving specifications and models of tractors produced including the Heider Model C 12-20 Hp.

Figure 6-57: Heider Model C 12-20 Hp, 1915

In 1916, Ford Motor Company began production of the Fordson tractor and crawler tractor, Model R, was marketed by Cleveland Motor Plough Co., Cleveland, OH. In 1917, Ford introduced the 20 hp Fordson tractor produced for the general trade by Henry Ford & Sons Co., Detroit, MI (spinoff of the Ford Motor Co.), weighing as little as one ton and advertised to sell for as little as $395. The Fordson soon ruled the tractor roost, accounting for 75 percent of the U.S. market share and 50 percent of the worldwide share. Nevertheless, the tractor business remained a competitive field, at least for a few decades, and competition helped foster innovations.

Figure 6-58: Fordson Model F tractor built in the UK from 1917 to 1964

Several tractor features such as mechanical lift for mounting and using attached equipment were built by Emerson- Brantingham, Ltd. in England. Standard belt pulley specifications, Power take-off (PTO) as optional equipment etc was introduced in tractor models in 1917. In the same year, Moline Universal tractor manufactured by the Moline Plough Co., was tested at the Columbus, OH trials; believed to be the first tractor tested with battery, starter, generator, lights and a rheostat control of engine speed by changing the voltage on the field windings of the generator.

Figure 6-59: Moline Universal, 1917

At about 1918, the Waterloo engine company used an inline two-cylinder engine which would be the type used by John Deere for the next 40 years to produce the tractor called the Waterloo Boy (See section on 'the John Deere revolution')

Figure 6-60: Waterloo boy, 1918

In 1919, the Nebraska Tractor Test Law, known as the Nebraska Inspection Law, went into effect and accepted as the basis for performance tests by industries, states, and other countries; SAE and ASAE involved. In 1920 the first official tractor test by the Nebraska Tractor Testing Laboratory (NTTL) was performed on the 26 hp Waterloo Boy N (NTT 1). 10-20 hp Titan tractor was the first Titan tractor tested in 1920 by Nebraska Tractor Testing Laboratory (NTT 23). Other tests were carried out on articulated tractors.

Same year, Starter and lights were made available for tractors and in 1921 use of tool bar for mounting implements on tractor began while lead was added to gasoline to reduce knocking in internal combustion engines, developed by Charles F. Kettering (1876-1958), and Thomas Midgley (1889-1944). The Rogers articulated 4WD tractor manufactured by Rogers Tractor and Trailer Co., Albion, PA., was tested (NTT 84) and found to produced 34.85 hp, with power steering.

Figure 6-61: Case 12-20 Tractor, 1922

In 1922, Wilson 4WD tractor, 12-24 hp, was developed by Wilson Brothers and marketed through the Wilson Tractor Manufacturing Co., Ottumwa, IA. In the same year, International Harvester Co. introduced the modern power take-off in its tractors and Case Co. also marketed her 12-20 tractor with top speed around 3 mph.

Case Co. and George White & Son, Ltd., Ontario, Canada stopped manufacturing steam engine tractors in 1924. Over the years of production Case built 35,737 steam engine tractors. In 1924 International Harvester Company introduced the Farmall tractor, one of the most successful models ever built.

Figure 6-62: International Harvester Farmall Regular, 1924

It was the first general-purpose tractor, designed for cultivating, ploughing and mowing. It was attractive to the average farmer and could work in a great range of crops. Soon other United Sates companies followed the trend and the general-purpose tractors replaced the one purpose tractors like the Fordson.

Also in 1924, mounted-type tractor implements were introduced and widely accepted and the 22-44 (dhp.-bhp[19]) tractors by Minneapolis Threshing Machine Co. the first commercial tractor with live PTO were built by Cockshutt Plough Co. Ltd., Canada. International Harvester Co., successfully introduced commercial 20 hp all-purpose row crop Farmall tricycle-type tractors, which was first tested in 1925 (NTT 117).

Figure 6-63: Farmall tricycle-type tractor, 1925

The Regular Farmall tractors were introduced in 1932 after the F-20. Some reasons for the success of the Farmall model were International Harvester's marketing networks, proximity to the farmers, knowledge of their customers' needs, credibility and a range of implements and accessories compatible with the tractor.

Figure 6-64: Farmall regular-type tractors, 1925

---

[19] dhp-bhp: drawbar horsepower-belt horsepower

John Deere GP (general purpose) Model D farm tractor, equipped with mechanical power lift to enable operator to raise and lower ploughs, discs, cultivators, and other attachments was introduced in 1928, Allis-Chalmers Co. manufactured the 6-cylinder Rumely tractor in 1930, then took over the Advance-Rumely Co, LaPorte, IN, in 1931. The era of incremental change set by the development of the Farmall tractors had many improvements in the performance and cost of tractors.

The turning point in tractor development

By the end of the 40s, tractors had settled into a common 'dominant design,' but there were still some technical differences and modifications yet to be made. By the end of the decade, the tractors that sold best in the U.S. looked a lot like the Allis-Chalmers Model "B." The turning point in tractor development was in 1945; when the amount of tractor power overtook the amount of horse power on American farms. The graph below indicated that one tractor could replace five horses on the farm.

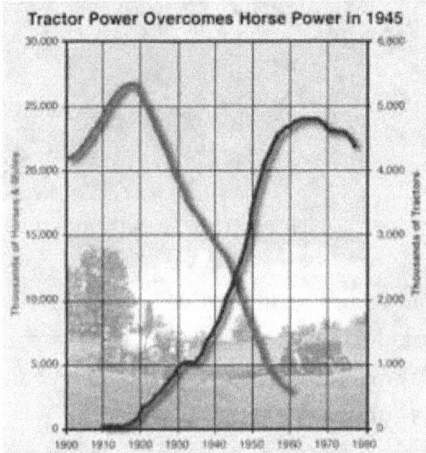

Figure 6-65: Animal and tractor power data from US Dept of Commerce

Tractors themselves got smaller and more lightweight and were designed with a higher ground clearance, making them capable of such relatively refined tasks as hauling cultivating implements through a standing crop. Though the design and reliability of tractors improved over the next 40 years, the tractor was still a mechanical 'draft animal', pulling ploughs, cultivators, or other equipment.

The tractor revolution and commercial sales continued until it peaked in 1951, when some 800,000 tractors were sold in the United States—equally important developments were occurring on the other side of the hitch. About that same time, the number of tractors began rising and peaked at just under 5 million in the late 60s and 70s. Between these periods, 4WD tractors were introduced and used on a large scale. Several 4WD tractors previously introduced but not used in large numbers are included in the timeline table below

Year	Tractor
1919	Fitch Four Drive, Sampson iron horse,
1922	Wilson four wheel drive,
1926	Wizard 4 Pull,
1929	Fitch Four Drive E,
1930	Massey Harris four wheel drive,
1936	Massey Harris four wheel drive,
1949	Detroit, Dodge four wheel drive power wagon,
1950	Harris power horse, general power built,
1953	Willis Farm Jeep,
1957	Mercedes Benz Unimog, Wagner TR9,
1960	Land Rover 88
1962	International Harvester 4300
1964	Case 1200,
1971	John Deere 7020, Minneapolis Moline A4T 1600, Steiger Bearcat and Massey Ferguson 1500
1975	White 4180
1976	Allis Chalmers 7580,
1978	Versatile 875, Ford FW30
1979	Belarus 1500,
1981	Big Bud 524/50

In 1965 Kubota, the first Japanese-built tractors tested at NTTL, available for gasoline or diesel fuel, produced by Kubota Iron and Machinery Works, Osaka, Japan (NTT 906).

6.3 Historic innovations in tractor development

Early innovations in tractor development

Power take off system (PTO)

One of the critical innovations, introduced early by International Harvester Co. in 1922, was the so-called power takeoff system where implements could be attached so they got their power from the tractor rather than wheels turning on the implement. This device consisted of a metal shaft that transmitted the engine power directly to a towed implement such as a reaper through a universal joint or similar mechanism; in other words, the implement "took off" power from the tractor engine. The John Deere Company followed in 1927 with a power lift that raised and lowered hitched implements at the end of each row — a time- and labor-saving breakthrough.

Diesel engine system

The first internal combustion tractors, starting with the Froehlich, used gasoline or other fuels like petrol or kerosene. Internal combustion engines powered either by gasoline or diesel fuel that had enough horsepower to do most jobs on the farm. Some tractors were equipped with all-fuel engines that allowed the farmer to use the cheapest fuel at the moment. Engines used different spark ignition systems from which the magneto became the most popular. And ever mindful of the power plant, at the same time the Otto-cycle engine was being developed another kind of internal combustion engine known as Diesel engine, which provided more power at a lower cost appeared.

In 1892 Rudolf Diesel patented an engine where fuel ignition was made by pressurizing the air until the temperature would ignite the fuel. The first engines tested used coal dust, but proved unpractical. Oil however proved to be a good fuel and the system could work with a four-stroke engine.

Benz developed the Benz-Diesel In 1922. One of the first companies to build diesel engines was Deutz that equipped many other German tractor constructors. Tractors equipped with diesel engines first appeared in Germany and from the 1930s on, diesel became more and more common in tractors in Europe, but in the United States they were mainly used in large crawler tractors and only became widespread after the 1970s. Since then almost every tractor model in the world is equipped with a diesel engine and gasoline is only used in small lawn tractors. Further developments in the diesel engine mostly depended on fuel injection technology.

Hydraulic system for lifting implements

The three-point hitch system for 'carrying' implements was first designed in 1928. By early 1930s, 1935 to be precise, an Irish engineer named Harry Ferguson had designed and already patented a the hydraulic draft control system for agricultural tractors, greatly improving the operator's ability to control implements, a duplex hitch and a draft control system which Ford introduced just before 1940 and had been adopted worldwide. Before this time, implements were dragged behind the tractor. By the next decade, all major tractor brands had their own versions.

Ferguson's system was revolutionary and greatly improved traction by transferring the implement's weight onto the back wheels of the tractor. The three-point hitch with hydraulic control allowed the farmer to lift implements at the touch of a button rather than brute strength on a large lever.

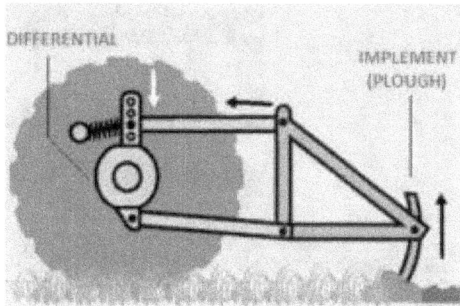

Figure 6-66: Side view of the Fergusson system (Ford archives, 2006)

The first Ferguson prototype had two upper hydraulic arms and the third hitch down, forming a V-shape. Soon it was turned upside down to the characteristic A-shape (Figure 6-66). The two hydraulic arms could control depth and if the attached plough would hit an object it could be lifted easily, preventing the tractor to topple backwards (Ferguson Family Museum, 2006). The first tractor to be commercialized with the complete three-point hitch was the Ferguson Black Tractor in 1933, manufactured in Northern Ireland. Though Ferguson's hydraulic system greatly improved tractor performance, it did not improve operator safety.

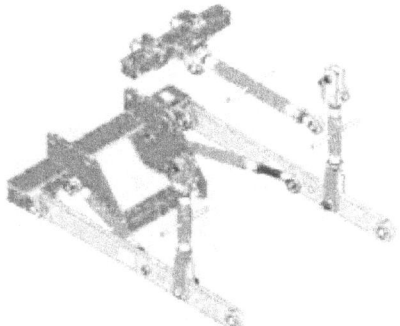

Figure 6-67: Schematic drawing of the Fergusson system (Ford archives, 2006)

Gasoline-powered track traction engine development

Early tractors were equipped with steel wheels. These had to be very large in the rear to support the weight, which made it hard to turn the tractor. Steel wheels were equipped with spikes that were capable to provide good traction in the field, but were very damaging to paved roads. Spikes were also damaging to superficial tree roots. The alternatives to steel wheels were crawler tracks.

The first built and patent for endless-chain traction engine was issued to Charles Dinsmoor, of Warren, PA, in 1886, but did not become a reality until 1904 by Benjamin Holt (1849-1920) who made the first agricultural traction engine with crawler tracks, which soon became known as the caterpillar.

Figure 6-68: Steam powered caterpillar tractor built by Holt in 1904

By 1904, self-laying track started to replace wheels for steam traction engines, by the Stockton Wheel Co. (which became the Caterpillar Tractor Co., Peoria, IL, in 1925), based on the work of Benjamin Holt (1849-1920), Daniel Best (1838-1923) and associates. Some of the rights for self-laying tracks by David Roberts (UK) were sold to Benjamin Holt (USA) who built crawler traction engines later in 1906. In 1908, Holt Co. developed its first commercialized crawler (tracklaying) tractor powered with an internal combustion (gasoline) engine.

At about 1912, double cylinder Gaar - Scott steam tractor & Cockshutt ploughs working in Saskatchewan was introduced.

Figure 6-69: Gaar - Scott steam tractor & Cockshutt plough, 1912

By 1925, merger between Holt Manufacturing Co., Daniel Best Company (formed by Daniel Best, (1838-1923), and C. L Best Gas Traction Co. (formed by Clarence L. Best (1875-1951)) formed the Caterpillar Tractor Co. in Peoria, IL. The wars of the 1920's lead this company to focus on the construction market, manufacturing crawler tractors, and left the agricultural tractor industry. Today crawler tractors are still more popular in construction than in agriculture.

In 1927 Deutz diesel tractors began commercial production in Cologne, Germany and ASAE (now ASABE) adopted its first standard for tractor PTO, drafted by the Farm Equipment Institute in 1926, with a speed of 536 ± 20 rpm. The standard has been updated several times since.

Figure 6-70: Early caterpillar type tractors (left; with crank start), 1920s

In the first half of the 20th century – before the great migration from the farm to the city first recorded in America in 1930, millions of people performed various kinds of farm works but few did that work on a crawler. In 1931, Diesel-powered track-type tractors were seen marketed by Caterpillar Tractor Co., and Caterpillar Diesel Model 65 was tested by Nebraska Tractor Testing Laboratory in 1932 (NTT 208).

Figure 6-71: Caterpillar tractor

Three important tractor features were introduced in the 1930´s: tyres, the Diesel engine and the three point hydraulic hitch system invented by Irish tractor maker Ferguson (Duarte and Sarkar, 2009). As tyres were broadly adopted to replace iron wheels, Ferguson's system was patented and only available on this make until 1938, when a gentlemen's agreement with Henry Ford allowed Fordson tractors to use the system (Brock, 2004).

Figure 6-72: Fordson roadless tractor

Other older tractors

During the 20th century, several tractors were designed with special features by different manufacturers. Some of the older tractors were as shown in Figures 6-73 below.

Figure 6-73: Older tractors

Rubber tyre revolution in tractor development

The history of tyres begins in 1844 with the development of rubber vulcanization by Charles Goodyear. In 1888, an Irish veterinary, J. C. Dunlop adapted a rubber tube to the wheels of his son's tricycle. This innovation marked the invention of tyre in vehicle transport. The Michelin brothers in France developed it further to be easier to change when damaged. In the United States, Henry Ford promptly adapted the new invention to his automobiles.

Figure 6-74: Henry Ford with his automobile

Rubber tyres designed for agricultural use came along in 1930s, making it much easier for tractors to function even on the roughest, muddiest ground. 1931 marked the beginning of rubber tyre revolution in tractor development history. Solid rubber tyre was invented by B. F. Goodrich Co. for farm tractors, which was soon replaced by low pressure pneumatic rubber tyres made by Firestone Tyre and Rubber Co.

Figure 6-75: Muri hill solid rubber tyre tractor

In 1932 Allis-Chalmers began to use pneumatic tyres from Firestone Tyre and Rubber Company; the Allis-Chalmers Model U tractor introduced in 1932 with rubber tyres had a 25 percent increase in fuel economy over steel wheels. The tires had a much better grip in the soils. They have many advantages over the metal tyres, including their weight.

Figure 6-76: Allis-Chalmers pneumatic tractor, 1932

By 1933, pneumatic rubber tyres were available on certain tractor models of the leading companies including Allis-Chalmers, Case, Deere, Ford, Huber, International Harvester, Massey-Harris, Minneapolis-Moline, and Oliver.

Figure 6-77: Pneumatic tyre tractors (left; Case tyre tractor), c. 1933

Same year, hydraulic system for lifting, adjusting depth of implements and for draft control was developed and introduced by Irishman Harry Ferguson (1884-1960). It was first built in England by David Brown Co., and manufactured in USA (1939-1948) by Ford Motor Co. McCormick-Deering W-12 became the last tractor using kerosene to be tested at NTTL, (NTT 229).

Figure 6-78: Pneumatic tyre tractors (left; Ford) and (right; Minneapolis-Moline)

Before 1938, there were no large low pressure rubber tyres made for farm machinery and all agricultural machinery, except farm trucks, travelled on steel tyres. By 1940 practically all farm tractors came with rubber tyres; *large rubber tyres* on the rear driving the tractor with small steering wheels in front either widely spaced or angled together in the middle.

Figure 6-79: Older tractors

From the late 1800s to 1940, the story of tractor development is one of smaller size, more power and less expensive machines. For many historians, the result of that development was the introduction of the Allis-Chalmers Model "B" in 1937. It was designed to pull a one bottom plough, cultivate corn, had rubber tyres, and sold for under $500. The Allis-Chalmers "B" sold well. Throughout the 1940s, other manufacturers scurried to match the innovation of the Model "B" and to improve on it.

Figure 6-80: Allis-Chalmers Model "B" tractor, 1937

The years between 1934 and 1947 saw several tractors passed the Nebraska Tractor Testing Laboratory standards and the introduction of several features notable was the Ford-Ferguson's introduction on their tractor N9 of a hydraulically actuated hitching and draft system with three links for attaching implements, now known as the three-point hitch, and has become known as a classic tractor. Also the PTO and drawbar dimensions were standardized in 1944, and later modified. Also introduced in 1947 was tractor with independent PTO, continuous running, that operated when the clutch was released.

Steering and transmission systems design and development

Power steering systems began, first with a kit to be added to tractors, manufactured by Behlen Manufacturing Co., Columbus, NE (Walt Behlen) in 1952. This was rapidly followed by many manufacturers including power steering as standard equipment. In 1953, Deere & Co. produced 4- and 6-cylinder engine tractors, replacing manufacture of 2-cylinder engine tractors.

Also in 1953, Allis Chalmers WD 45 was the first tractor to have a snap coupler for a remote cylinder with control lever mounted on the steering pedestal (NTT 499). International Harvester Co. introduced the torque amplifier (TA), a major development in transmission design in 1955 and Hydrostatic transmission later in 1964.

Roll-over protective structure, ROPS

In 1956, Roll-over protective structure, ROPS, for tractors developed by Lloyd H. Lamoria, Ralph R. Parks, and Coby Lorenzen at the University of California, Davis, CA. was introduced and tested in California, First commercial ROPS offered by Deere & Co. in 1966 through the efforts of Charles Morrison on John Deere tractors; first on the JD 4020, following considerable research and joint work with industrial, educational and government organizations (NTT 934). National Safety Council approved tractor overturn protection system in 1972, certified as ROPS (roll-over protective structure). Around 1985, it became mandatory to fit tractors with 'roll over protection structures' to lessen the high risk of fatal injury if the tractor rolled over.

Tractor sound testing

By 1970, standards on tractor sound testing were performed by NTTL. The first tractor tested for sound was the 86 hp Case 970 Diesel tractor, using dB (A) decibel scale and followed in test procedure for tractors tested at the NTTL, as requested by OSHA of the U.S. Department of Labour. Beginning in 1971 all tractors tested for sound at the operator station for tractors with and without cabs, and at a position 7.5 meters to the side.

Other designed features

Electronic fuel system, which controlled the air and fuel of the engine for greater efficiency was developed in Great Britain in 1966. By 1986 electronic control of fuel injection systems for diesel engines for tractors introduced. The first satellite launched for Global Position System (GPS), from which many applications developed including steering of tractors and implements was in 1978.

John Deere tractor revolution

Following his huge successes in designing and manufacturing of agricultural implements, John Deere Co decided to get into the tractor business by buying out the Waterloo engine company in 1918, and 1923 built its first prototype tractor, the Model D. the company rapidly improved on its tractors, producing Model C, A, B and the general purpose. John Deere Co expanded its business in 1950 by putting up an 8 million dollar factory on 1,400 acres along the Mississippi river to manufacture combines, garden tractors, sprayers and other agricultural equipments.

Figure 6-81: John Deere tractors

In 1938, John Deere Co grew its business redesigning their tractors to be streamlined with no bolts or seams showing outside of the hood and better view from the seat. In 1960, John Deere produced 4- and 6-cylinder engines and 95 hp 4020 engine to pull a 5-bottom plough. In 1972, generation II was introduced, while in 1994 the 8000 series tractors were introduced with new technologies to meet the new emission standards.

Figure 6-82: John Deere tractors

6.4 Future of tractor development

Future of tractor development

After the 40s, as tractor manufacturers seek better productivity through improvements in basic design features, the machines got larger and more powerful but are becoming heavy again. To avoid soil compaction problems, modern tractors are sometimes '4-wheel driven' (4WD), with the weight evenly distributed over the four wheels. Dual or twin wheels are sometimes fitted to further reduce ground pressure. By 1986 Track-type tractor introduced with rubber tracks, the Caterpillar Challenger 65, with 270 engine hp and higher.

Figure 6-83: Today's tractor

However, with all the developments in the last hundred years, change has been slow not because tractor designers and engineers have been idle, but because manufacturers are reluctant to make expensive changes to their production lines. The future will see more sophisticated tractors with greater reliability, maneuverability, comfort, and safety. Gradual evolution in design is more likely than radical change. Currently, faster highway towing speeds and improved braking systems are being developed.

Tractor design and layout

Tractors are divided into six main types:

Two wheeled tractors: Two wheeled tractors are often called 'walking tractors'. The small hand held units with rotary cultivators are usually driven by petrol engines. Larger units are often coupled to 2 wheeled trailers and driven by diesel engines. Engines of 5-10 horsepower (HP) are common.

Four wheeled, 2WD tractors: Four wheeled, 2WD tractors are the most common type of tractor. The two rear wheels supply power. The front wheels are much smaller and are used only for steering. Engine power ranges from 25 to 120 HP.

Four wheeled, 4WD tractors: The 4 wheeled, 4WD tractor is similar to the 4 wheeled, 2WD tractors, but all 4 wheels can be powered for better traction. The front wheels are half the size of the rear wheels. 4WD tractors usually have power ratings of 70 HP and above, although some lower horsepower units are made for special purposes.

Eight wheeled, 8WD tractors: The 8WD tractors are the top of the range in terms of weight and power. They avoid soil compaction and wheel slip by distributing weight and power over 8 wheels. They are jointed in the middle, and 1 axle is mounted to each of the jointed halves, with both axles driven. Each axle has 2 dual wheels at each end, or 4 wheels per axle. 8WD tractors usually have engines of 200 HP or more.

Tracked or 'crawler' tractors: Tracked tractors are equipped with tracks instead of wheels. They are generally used on farms where soils are difficult to cultivate, such as heavy clays, or where seasons are shorter and wet conditions predominate. More power can be transmitted to the drawbar than wheeled tractors.

Maintenance costs are however higher than for wheeled tractors. Also, unlike wheeled tractors, they must be transported from field to field by a 'low loader'. Recently, a tractor manufacturer designed a high powered, high speed tracked tractor that runs on rubber tracks and can be driven on the road. This model may make the tracked tractor more popular in future. Tracked tractors have power ratings of 65-700 HP.

Special purpose tractors: The agricultural tractor is sometimes modified for use under special conditions where the standard type is not suitable.

The narrow tractor is a basic small tractor which has been modified to pass through the narrow rows of soft fruit trees and vineyards.

The high clearance tractor is usually a standard tractor which has been elevated to give more clearance over advanced stage crops for spraying or fertilizer broadcasting.

Light, low HP, 4WD tractors work in rice paddies, where the combination of light weight and 4WD is desirable.

Two-wheel tractor: The higher power "riding" rotavators cross out of the home garden category into farming category, especially in Asia, Africa and South America, capable of preparing 1 hectare of land in 8-10 hours. These are also known as *power tillers* or *walking tractors*. Years ago they were considered only useful for rice growing areas, where they were fitted with steel cage-wheels for traction, but now the same are being used in both wetland and dryland farming all over the world. They have multiple functions with related tools for dryland or paddys, pumping, transportation, threshing, ditching, spraying pesticide. They can be used on hills, mountains, in greenhouses and orchards. Diesel designs are more popular in developing countries than gasoline.

Figure 6-84: Walking tractor (power tiller)

Other special purpose tractors provide engine and transmission facilities for purpose-built machines, such as pipe layers, drainage machines, and mechanical harvesters.

6.5   Most significant innovations in farm power development

Inventions considered most significant in the development of farm power include those of:

1. **Horse collar**: The horse collar, which allowed horses to pull much heavier loads, was invented in China long before it was known in Europe by the 9th century. Previously horses were attached to vehicles by straps around their necks. The strap would constrict its neck and could not allow the horse pull a heavy load until the collar was invented! This had made horses used more efficiently in the middle Ages.
2. **Steam powered and traction engine**: The more efficient steam powered traction engine, which was adapted to power machinery, was patented in 1769, by James Watt. In the 19th century machines in factories were seen operated by steam engines. At the end of the 19th century they began to convert to electricity. In 1887, steam-powered traction engine (later called tractor) was patented and manufactured by Daniel Best (1838-1923), and also invented a combined steam-driven harvester and thresher.
3. **Internal combustion engine**: The internal combustion engine which was able to solve the risk of boiler explosion and more power development in engine and ultimately in 1889, gas(oline)-powered internal combustion traction engine appeared and was first built by Charter Gas Engines Co., Chicago, IL (patented by John Charter in 1887) with the power unit built on a Rumely steam traction engine frame, called a Burger traction engine, believed to be named after the developer, but the unit could not pull single plough. In 1892 One of first practical gasoline traction engines (20 hp) that was an operating success; single cylinder, included a clutch, and could be propelled forward and backward was built by Van Duze Gas and Gasoline Engine Co., OH, for John

Froelich in Iowa; the forerunner of the Waterloo Boy traction engine later manufactured by John Deere & Co.

Also in the same year, design and drawings for first gas(oline) traction engines for J. I. Case Threshing Machine Co. by David Pryce Davies (1870-1948), was first used to power a thresher. At the time J. I. Case was the largest manufacturer of steam engines in the world. Another engine built to ignite fuel by heat of compression was patented by Rudolph Diesel (1858-1913), first designed to use powdered coal but later used liquid fuels.

4. **Tractor**: The traction vehicle which was later known as the tractor powered the agricultural mechanization revolution.

CHAPTER 7

AGRICULTURAL MACHINERY
Historic Developments

7.0 Introduction

Farm machinery has continued to evolve from the Old Stone Age into the world of emerging and sustainable technologies. Today's farm machinery input has allowed farmers to cultivate many more acres of land than the machines of yesterday. The old hunting and gathering society had given way to primitive old artifacts of wood and stone tools had been replaced with the more durable and stronger traditional agricultural tools made from iron which had revolutionized mechanized agriculture with more sophisticated machinery equipped with GPRS systems as it now exists today.

Mechanization, the outstanding feature of agriculture in the late nineteenth and twentieth century has relieved much of the farmer's work. The development of high tech farm equipment implies that fewer people will be directly involved in farming but many more are involved in the agricultural industry from scientists to salespeople. The transmission of vital weather data and other information of interest to farmers through radio and television have significantly increased efficiency and productivity of farms.

These sophistications and improved farming systems had made the ploughs not to be extensively usable as before, largely due in part to the popularity of minimum tillage to reduce soil erosion and conserve moisture as well as the soil carbon content. The disk harrow today is more often used after harvesting to cut up the grain stubble left in the field. Although seed drills are still used, the air seeder is becoming more popular with farmers. The threshing machine has given way to the combine, usually a self-propelled unit that either picks up windrowed grain or cuts and threshes it in one step. The grain binder has been replaced by the swather that cuts the grain and lays it on the ground in windrows, allowing it to dry before being harvested by a combine.

The transition from horse drawn, simple basic equipment to the modern high tech equipment of today has actually happened in a farmer's life span. People working in research and development are constantly looking for ways of improving farming equipment and farm practices. Planes and helicopters are now used for agricultural purposes, such as planting, transportation of perishable goods and fighting forest fires and crop fumigant to control insect pests and diseases. Producing, marketing, and maintaining this high-tech equipment requires precise metal workers, highly skilled technicians, and knowledgeable sales personnel.

From the foregoing, much has changed from the earliest wheat gatherers of 9000 years ago to farmers of today, but the biggest changes are actually very recent as discussed in the following sections.

7.1   History of land clearing and development

Since the end of the 1800's, grasslands have diminished in quality and quantity due to changes in agricultural practices, increased farmers activities, and an increase in human population. In most part of the world, land clearing has been utilized to make way for agricultural and urban development. In the past, governments and people thought that if land was left on its own that it was being "wasted" when it could be put to good use to be developed for agricultural purposes.

By taking scrub land, clearing it, and turning it into fields for crop production not only was the increase in land value raised, but so was economic gain for the community. While at one time land clearing was seen as beneficial and even progressive, it is now generally viewed as destructive.

Figure 7-1: Land clearing by farm settlers (Paterson Global Foods Inc)

Since more environmental awareness has taken hold, countries which use land clearing keep legislative regulation on its use. Despite the known negative environmental impact, farmers worldwide object to the restriction of land clearance because it effects their crop production and how much land they have available to them.

Land clearing defects

Deforestation: The Neolithic period saw extensive deforestation for farming land (*Encyclopædia Britannica Online*). With the advent of agriculture, larger areas began to be deforested, and fire became the prime tool to clear land for crops. The first evidence of deforestation appears in the Mesolithic period (Tony, 1997). It was probably used to convert closed forests into more open ecosystems favourable to game animals (Flannery, 1994). Small scale deforestation was practiced by some societies for tens of thousands of years before the beginnings of civilization (Flannery, 1994).

Mesolithic foragers used fire to create openings for red deer and wild boar. In Great Britain, shade-tolerant species such as oak and ash are replaced in the pollen record by hazels, brambles, grasses and nettles. Removal of the forests led to decreased transpiration, resulting in the formation of upland peat bogs. Widespread decrease in elm pollen across Europe between 8400-8300 BC and 7200-7000 BC, starting in southern Europe and gradually moving north to Great Britain, may represent land clearing by fire at the onset of Neolithic agriculture.

Figure 7-2: Deforestation of Brazil's Atlantic Forest c.1820-1825

In Europe there were little solid evidences of land clearing activities before 7000 BC. Evidence of deforestation has been found in Minoan Crete; for example the environs of the Palace of Knossos were severely deforested in the Bronze Age (Hogan, 2007).

From 1100 to 1500 AD, significant deforestation took place in Western Europe as a result of the expanding human population. The large-scale building of wooden sailing ships by European (coastal) naval owners since the 15th century for exploration, colonization, slave trade–and other trade on the high seas consumed many forest resources. Piracy also contributed to the over harvesting of forests, as in Spain. This led to a weakening of the domestic economy after Columbus' discovery of America, as the economy became dependent on colonial activities (plundering, mining, cattle, plantations, trade, etc.)

Erosion: In ancient Greece, Tjeered *et al.*, 2007, summarized three regional studies of historic erosion and alluviation and found that, wherever adequate evidence exists, a major phase of erosion follows, by about 500-1,000 years the introduction of farming in the various regions of Greece, ranging from the later Neolithic to the Early Bronze Age.

7.2 History of farm machinery development

Modern farmer needs a wide range of tools and machines for loosening the soil, getting rid of weeds, planting, mowing, and harvesting the crops. There are specific machines designed for the different crops that are grown on farms. Some are designed for grain crops, while others are designed to plant and harvest potatoes, lentils, and other crops. The long transformation from simple equipment, like the sickle and the horse-drawn plough, to the modern high tech equipment which has changed farming today started at about 70 AD when Elder Pliny reported to Roman leaders the use of a wooden cart with comb-like bars pushed by animals for reaping wheat.

In the 1st century A. D., perhaps a monument qualified to be the world's first agricultural engineers was found in Gaul, with a painting showing a harnessed mule pushing a wooden harvester through a field of wheat, a model of which is at ASABE headquarters, courtesy of Wayne Worthington.

Inventions in the ancient world

Sometimes after the fall of the Roman Empire in the 5th century AD, there was a dark age in Europe. However a number of inventions were made across Europe at that time or reached the continent from other cultures. Notable in the field of agriculture and mechanization include the followings.

Crop rotation

In the Ancient World, land was divided into 2 fields, one of which was sown while the other was left fallow. One big improvement in agriculture was the 3-field system invented in Germany in the 8th century. One field was sown in Spring, one in Autumn and one was left fallow. This system allowed farmers to grow more food.

However, in the 17th century, new forms of crop rotation were introduced. The Dutch began to use new forms of crop rotation with clover and root crops such as turnips and Swedes instead of letting the land

Figure 7-3: Charles Townshend

grow fallow. (Root crops restored fertility to the soil). In the 18th century these new methods became common in England. A man named Charles 'Turnip' Townshend (1674-1738) did much to popularize growing turnips.

History and development of plough

The plough (BrE) or plow (AmE)[20] is a tool (or machine) used in farming for initial (primary) cultivation (working) to loosen or turn the soil in preparation for seed sowing or planting. The primary purpose of ploughing is to turn over the upper layer of the soil, bringing fresh nutrients to the surface, while burying weeds, the remains of previous crops, and both crop and weed seeds, allowing them to break down.

The plough was traditionally known by other names, e.g. *sulh* in Old English, *medela, geiza, huohili* in Old High German language and *arðr* in Old Norse (Swedish *årder*), all presumably referring to the scratch plough. The current word *plough* comes from Old Norse *plógr*. The plough was the most important piece of equipment on a farm during the 19th century; used to break up and turn soil for planting crops, a whole day was required to plough 1-2 acres of land with a mouldboard plough.

Generally, a farmer who owned a plough would hire it out to those in need, in exchange for labour or goods. Oftentimes a farmer would not personally own a plough as they were considered to be expensive. Occasionally, a community would pool resources to purchase a plough.

Early ploughs

The earliest forms of ploughs were made of forked sticks and timbers. In the Middle East the early ploughs were called ard. The early ploughs simply loosened the soil. A type of ard is still seen and used in some underdeveloped countries today.

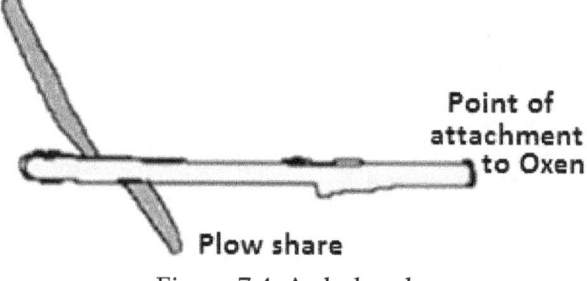

Figure 7-4: Ard plough

---

[20] BrE ⇒ British English; AmE ⇒ American English

Human powered plough

Before the invention of plough somewhere around 6,000 BC, man cultivated land with hoe like devices. Initially ploughs were powered by humans, but the process became considerably more efficient and progressive once animals were pressed into the service.

Breast ploughs

The age and origin of the breast plough is unknown. Some have speculated on its possible use in Roman times, that it may have been a development of the Parsnip shovel illustrated in Markhams "Farewell to Husbandry" of 1620 or the double flanged trenching spade shown in Blith's "English Improver Improved" (Edilson, 1653), but there is little or no evidence to support these theories.

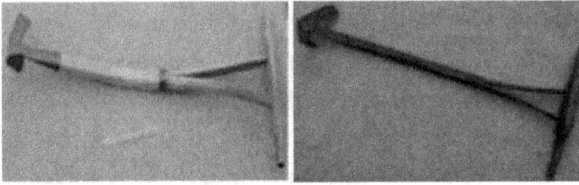

Figure 7-5: Breast ploughs

Taken with other contemporary references it may be reasonably assumed that the breast plough was in use by 1650 and continued in service until around 1850 when this method of paring rapidly declined. After that it remained in sporadic use in certain areas such as the Cotswolds until the 1930's or 40's when it was employed latterly as a garden, allotment or small holding's tool often by the men who had previously worked them across old pastures.

The word breast also suggests that the plough was propelled from the chest rather than the haunches. Working in rows from left to right with the handle held palms uppermost, the feet splayed and the knees bent, the 'plough' was shoved forwards from the upper thighs in a series of jerks which sliced off a thin layer of turf or topsoil in lengths of one to three feet (310-914 mm) to a depth of one to three inches (25-76 mm). Then, with a dexterous twist of the handle it was overturned or 'whelmed over' from left to right opposite the flange (or vice versa if a right-handed flange) depositing the cut sod face down on the ground to weather and dry.

Figure 7-6: Breast plough courtesy of Roy Brigden

Breast ploughs varied in size and shape from region to region adapted to suit local conditions and methods of construction and not least made to measure requirements of users. Though semi-circular, crescent or intermediate blade patterns were produced, the majority were of triangular form made by blacksmiths from a single piece of iron or steel plate approximately 1.6 to 6 mm thick, between 203-458 mm in width and up to 432 mm in length with a sharply honed soft or deep 'V' cutting edge, the point being set in line with the shaft.

Uses: Breast ploughs are primarily tools for small farmers used for paring stubble and weeds, leveling mole or anthills, ploughing in potatoes, or vetches, work on water meadows and in some areas, cutting peat for fuel, cleaning roadside gutters and overgrown grass verges.

Ploughing with animals

In Mesopotamia (today's Iraq) and Indus Valley (Pakistan-India) man first harnessed the ox and so the plough. These animals appear to be first used around 3500 B.C. with primitive ploughs made of wood. The first animal powered ploughs were pulled by oxen (any kind of cattle used for draft, or pulling work, are called oxen), and later in many areas by horses (generally draught horses) and mules, although various other animals have been used for this purpose (Halleym, 1996).

Figure 7-7: Ploughing with oxen

The domestication of oxen in Mesopotamia and by its contemporary Indus valley civilization, perhaps as early as the 6th millennium BC., provided mankind with the draft power necessary to develop the larger, animal-drawn true **ard** (or scratch plough). The earliest was the *bow ard*, which consists of a *draft-pole* (or *beam*) pierced by a thinner vertical pointed stick called the *head* (or *body*), with one end being the *stilt* (handle) and the other a *share* (cutting blade) that was dragged through the topsoil to cut a shallow furrow ideal for most cereal crops.

Figure 7-8: Ancient Egyptian ard, c. 1200 B C

The ard does not clear new land well, so hoes or mattocks must be used to pull up grass and undergrowth, and a hand-held, coulter-like *ristle* could be used to cut deeper furrows ahead of the share. Because the *ard* leaves a strip of undisturbed earth between the furrows, the fields are often cross-ploughed lengthwise and across, and this tends to form squarish fields (Celtic fields) (Lynn, 1962).

The *ard* is best suited to loamy or sandy soils that are naturally fertilized by annual flooding, as in the Nile Delta and Fertile Crescent, and to a lesser extent any other cereal-growing region with light or thin soil. By the late Iron Age, ards in Europe were commonly fitted with coulters. Some ancient hoes, like the Egyptian ard, were pointed and strong enough to clear rocky soil and make seed drills, which is why they are called hand-ards.

Figure 7-9: Animal drawn ard plough

Animal drawn ard plough as shown in Figure 7-9 were used for primary tillage and weed control. The invention of the horse collar and shoe in Europe in the 9th century allowed the plough to be pulled by horses, yet even well into the 18th century, oxen still outnumbered the horses partly due to the expenses on feeding the horses (Grigg, 1974). However, with the advent of the iron ploughs, many farmers changed over to the use of heavy horses which could pull the new type of implements at a faster pace than the oxen.

Figure 7-10: Traditional way of horse ploughing

Engine powered plough

The advent of the mobile steam engine allowed steam power to be applied to ploughing from about 1850. In industrialized countries, the first mechanical means of pulling a plough were steam-powered (ploughing engines or steam tractors), but these were gradually superseded by internal-combustion-powered tractors. In Europe, soil conditions were often too soft to support the weight of heavy traction engines. Instead, counterbalanced, wheeled ploughs, known as *balance ploughs*, were drawn by cables across the fields by pairs of ploughing engines on opposite field edges.

The man credited with the invention of the ploughing engine and the associated balance plough, in the mid nineteenth century, was John Fowler, an English agricultural engineer and inventor. Such ploughing engine known as *Heumar*, was built in 1929, by the Ottomayer Company, Germany (shown in Figure 7-11) used in pairs with a balance plough and weighing 220 *lbs*, (21 tons).

Figure 7-11: Ottomayer ploughing engine *Heumar*

Heavy ploughs

To grow crops regularly in less fertile areas, the soil must be turned to bring nutrients to the surface. A major advance for this type of farming was the *mouldboard plough*, which not only cuts furrows with a share (cutting blade) but turns the soil. The origin of the invention of mouldboard (American spelling: *moldboard plough*; or turnplough, frame-plough) is contentious, however, history recorded that by 100 BC, iron moldboards were used in Chinese ploughs and the earliest iron ploughshares from ca. 500 BC in China (Greenberger, 2006).

Figure 7-11: Chinese iron plough with curved mouldboard, 1637

Before the Han Dynasty (202 BC–220 AD) era, Chinese ploughs were made almost entirely of wood, except the iron blade of the ploughshare. By the Han's period, the entire ploughshare was made of cast iron; these are the first known heavy mouldboard iron ploughs (White, 1984; Zhongshu, 1982). By 100 BC, iron moldboards were used in Chinese ploughs and by 1000 AD; the Dutch farmers had copied or also designed them.

Figure 7-12: Han Dynasty mural, depicting ploughing in Shennong

Virgil (Publius Vergilius Maro) wrote around 1 AD about the Roman plough with an iron ploughshare......."From its youth up, in the woods, the elm is bent by main force and trained for a plough stock, taking the form of a crooked plough: to suit this a beam is shaped stretching eight feet in front, while behind are attached two mold boards resting on the slade (or sole piece) with a double ridge".

The major advance before 1000 BC was the development of the heavy plough, which was more than the simple ploughs that farmers used earlier. The earliest ploughs with a detachable and replaceable share dated from around 1000 BC in the ancient Near East (White, 1984) which the Dutch farmers had copied or also designed about same time. Mouldboards are known in Britain from the late 6th century on. Early mouldboards were basically wedges that sat inside the cut formed by the coulter, turning over the soil to the side.

Figure 7-13: Early (Carey) plough

The first indisputable appearance of mouldboard in a northern Italian document after the Roman period was from 643. Old words connected with the heavy plough and its use appeared in Slavic, suggesting possible early use in this region. The general adoption of the mouldboard plough in Europe appears to have accompanied the adoption of the three-field system in the later 8th and early 9th centuries, leading to an improvement of the agricultural productivity per unit of land in northern Europe (White, *Medieval Technology*).

Research by the French historian Marc Bloch in medieval French agricultural history showed the existence of names for two different ploughs, "the ard (*araire*) was wheeless and had to be dragged across the fields, while the turnplough (*charrue*) was mounted on wheels (Bloch, 1966)

Early walking ploughs (also known as wheelless plough)

In 1653, the first known treatise on plough construction, *The English Improver Improved*, by Walter Blith in England, was printed for J. Wright. Moldboard for a plough based on scientific principles was invented (but not patented) by Thomas Jefferson (1743-1826), of Virginia in 1703. Plough developments preceded this invention in England with patents to Joseph Foljambe in 1720. The first English patent granted on a plough was to Joseph Foljambe, of Yorkshire, in 1720, he having invented a number of improvements on a crude style of plough, which had been brought from Holland. The bottom of Foljambe's plough was of wood, with a sheet-iron covering on the wearing parts and a point of iron plate. The coulter was, of course, made of iron. The point was conical in form and the furrow was raised by it and then turned over by the mouldboard.

But this plough, although it was superior to anything then known, did not come into general use until James Small established his factory at Black Alder Mount, Scotland, in 1763, and began to manufacture and sell ploughs on what was then a large scale. In time he made many improvements, and the plough finally assumed the style of the East Lothian, which gave general form and feature to all the common British ploughs since.

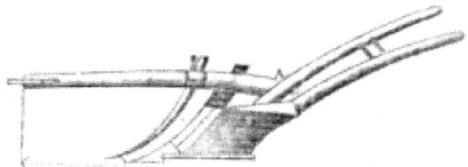

Figure 7-14: Drawing of Rotherham plough

Joseph Foljambe's Rotherham cast iron plough of 1730, combining an earlier Dutch design with a number of technological innovations, was the first iron plough to have any commercial success in Europe. Rotherham used new shapes as the basis for the Rotherham plough, which also covered the mouldboard with iron. It was much lighter than conventional designs and became very popular in England. It may have been the first plough that was widely built in factories.

The Rotherham plough measured 7 feet, 4 inches from the end of either handle to the point of the share. The length of beam is 6 feet. Length of the landside and share measures 2 feet, 101 inches as they run on the ground. The height from the ground to the top of the beam where the coulter goes through is 1 foot, 8 inches while the weight of wood and iron work total 140 lbs. Its fittings and coulter were made of iron and the mouldboard and share were covered with an iron plate making it lighter to pull and more controllable than previous ploughs. It remained in limited use in Britain until the development of the tractor.

Figure 7-15: Rotherham plough

In 1740, plough with cast iron moldboard and wrought iron ploughshares was invented by Scot James Small of Doncaster at his Blackadder Works. James Small and Berwickshire further improved the design in 1763, using mathematical methods he experimented with various designs until he arrived at a shape cast from a single piece of iron, the *Scots plough*. "Scots Plough" used an improved cast iron share to turn the soil more effectively with less draft, wear, or strain on the ploughing team.

Figure 7-16: Old English plough

Most teams pulling ploughs converted to horse drawn ploughs sometime after 1850. Factories for making ploughs were established in England in 1783, and an iron plough was invented by Scot James Small, Berwick, England and tempered cast iron ploughshares were patented by Englishman Robert Ransome (1753-1830), which was practically built in 1785 as ploughs with detachable parts. In 1790, mole plough was developed by Mr. Vaisey in England.

The first letters patent granted in America, on a plough, one-piece cast iron plough, not including beam and handles, was in 1797, to Charles Newbold (1780-??), a farmer of

Burlington County, N. J., Scot James Smith in 1767, and Englishman Robert Ransome in 1785. This was again improved on by Jethro Wood, a blacksmith of Scipio, New York, who made a three-part Scots Plough that allowed a broken piece to be replaced. Between 1814 and 1819, cast iron ploughs with improved moldboard in which wearable parts could be replaced were invented and patented in 1819 by Jethro Wood (1774-1834), in Scipio, NY.

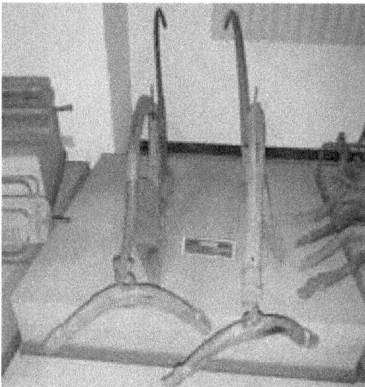

Figure 7-17: The Philippines ancient plough

Jethro Wood's invention, patented September 1, 1819, ushered in a new era in the history of the plough, the era of manufacturing, as distinguished from the era of building in small quantities by blacksmiths or "plough wrights". Credited, along with Jethro Wood, 1816 saw a further development in the introduction of cast ploughs with wrought iron point, invented and patented by Stephen McCormick (1784-1875), Auburn, VA., with introducing the cast iron plough in the USA.

Figure 7-18: Plough c.1800-1850

In 1831, side-hill plough was patented by Cyrus H. McCormick (1809-1884), and John Lane of Lockport, IL; a blacksmith first manufactured a steel plough with strips of steel over wood moldboards in 1833 but later, it was improved by his son John Lane (1824-1897), Chicago, IL., but no patency was issued until John Deere steel plough was separately patented.

Scot James Smith, of Deanston, Scotland developed and used subsoil plough that combined subsoiling and drainage in 1823. Joel Nourse was one of the rioted ploughmen of the generation succeeding Wood. He first started at Shrewsbury, Mass., but afterwards removed to Worcester, and in 1842, perfected the famous Eagle series, ploughs with a longer mouldboard than Wood's, and with a greater turn, breaking the furrow more thoroughly. The sales in the forties of Nourse's firm, (Ruggles, Nourse, Mason & Co.), were said to have reached 25,000 and 30,000 ploughs per year.

Figure 7-19: Norse eagle plough

The problem among others was solved by a black smith from Vermont named John Deere who moved to Grand Detour, Illinois in 1836. He invented a blade which was self polishing and combined the share and mouldboard into a one piece plough. The blade was an amazing hit; it was so much stronger than iron designs that it could work soil in areas of the US that had previously been considered unsuitable for farming.

Figure 7-20: John Deere self polishing plough

Improvements on this followed developments in metallurgy: steel coulters and shares with softer iron mouldboards to prevent breakage, the chilled plough (an early example of surface-hardened steel) (John Deere, 1804-1886) and eventually the face of the mouldboard grew strong enough to dispense with the coulter.

Figure 7-21: John Deere plough

By 1832, steam drawn plough, using steam engine and cable, was developed by John Heathcoat (1783-1861), and Henry Handley. Cable for ploughing was invented by E. C. Billinger of South Carolina in 1833. As early as 1837, steam plough (steam engine pulling a cable) was patented by John Upton in England.

Figure 7-22: Early ploughs

A significant invention in equipment development was the disc plough developed in Australia by John Shearer in 1847 and the revolving disk harrow (similar to disc plough) patented in USA by George Page same year. Disc plough market did not develop until 1893. John Fowler (1826-1864), developed and publicly exhibited the mole draining plough in 1851, used cable for ploughing in 1852 and used steam power applied by cable to draining plough in 1853, followed by its use for cultivation in 1855 and for other agricultural equipment made by equipment manufacturer, Ransomes, Ltd.

The invention of the chilled plough

There remains to be noticed an important step in the perfection of the ploughs in use throughout the eastern states. Efforts to harden the wearing parts, and thus make them more durable, began almost with the first use of cast ploughs, but the chilling process was so little understood, that for more than half a century no one could master it. Credit for making the chilled plough a practical success is due to James Oliver, who began experiments soon after establishing his plough shop or factory at South Bend, Ind., in 1853.

Chilled-iron was plough developed and patented in 1854 with an annealing process introduced in 1857. This implement became one of the most used ploughs in USA. Made by James Oliver (1823-1908) of the Oliver Chilled Plough Works. In 1857, a steam plough pulled by cable was invented by Richard Jordan Gatling (1818-1903), who was famous for developing the Gatling machine gun. In 1859, steam plough was patented and built in Pennsylvania by John W. Fawkes while steam ploughing apparatus was patented by Englishman John Fowler (1826-1864).

Many ploughs had wheels added to the side or back of the plough that allowed the plough to be held vertically; easier to be raised or lowered easily for transport to different fields. Without wheels the plough had to be either dragged on its side or loaded and hauled in a cart.

Figure 7-23: Pennsylvania wheel plough

Plough revolution witnessed the introduction of two-wheeled sulky plough in 1864, including seat for operator, patented by two Americans, F. S. Davenport and Robert Newton of Illinois, which was replaced by the Flying Dutchman, the three-wheeled sulky plough produced by Moline Plough Co., USA in 1884.

Disc plough, known as the Sovereign plough built by John Shearer and Sons in Australia, gained prominence in 1877 as Brantford Plough Works founded in Brantford, Canada, to build tillage implements by James G. Cockshutt (1853-1885) family same year. The Company named Cockshutt Plough Co. Ltd. In 1882 expanded into tractor production, building the first tractor with live PTO in 1924. By 1893 disc plough reappeared in USA after its first introduction in 1847, but not used at the time until 1896 when the first successful disc plough was invented in USA by Mr. Hardy.

After about 1800 all these metal ploughs were made in quantity and at "reasonable" prices because of the production of cheap steel and iron produced by the development of the steel making industry in Britain. Robert Ransome's plough factory of Ipswich, England was reported to have produced 86 different plough models designed for different soils (Barlow, 2003).

Tractor-carried ploughs was patented and produced in 1927 by manufacturers, one of which was the French Huguet-Huard and in 1935, Lift-type mounted ploughs had been introduced and had gained in availability. Heavy-duty chisel plough, which helped control wind erosion, was invented by Fred Hoeme, Hooker, OK in 1933

Riding and multiple-furrow ploughs

Wheels were added to the plough by different people, mostly in the Middle Ages with some indications that the Romans invented the heavy wheeled mouldboard plough in the late 3rd and 4th century AD, when archaeological evidences appeared, inter alia, in Roman Britain (Margaritis, 2008). The wheels allowed a deeper cutting of the soil.

Figure 7-24: Roman wheel plough

Early steel ploughs, like those for thousands of years prior, were *walking ploughs*, directed by the ploughman holding onto handles on either side of the plough. The steel ploughs were so much easier to draw through the soil that the constant adjustments of the blade to react to roots or clods were no longer necessary, as the plough could easily cut through them.

Figure 7-25: Early tractor drawn two-furrow plough

The first *riding ploughs* appeared not long after the walking ploughs were invented. The wheels kept the plough at an adjustable level above the ground, while the ploughman sat on a seat; whereas, with earlier ploughs the ploughman would have had to walk. Direction was now controlled mostly through the draught team, with levers allowing fine adjustments. This led very quickly to riding ploughs with multiple mouldboards, dramatically increasing ploughing performance.

Figure 7-26: John Deere's early wheel plough

A single draught horse can normally pull a single-furrow plough in clean light soil, but in heavier soils two horses is needed, one walking on the land and one in the furrow. For ploughs with two or more furrows more than two horses are needed and, usually, one or more horses have to walk on the loose ploughed sod—and that makes hard going for them, and the horse treads the newly ploughed land down. In heavy volcanic loam soils, such as are found in New Zealand, require the use of four heavy draught horses to pull a double-furrow plough. It is usual to rest such horses every half hour for about ten minutes.

Amish farmers tend to use a team of about seven horses or mules when spring ploughing. Using this method about 10 acres (40,000 m²) can be ploughed per day in light soils and about 2 acres (8,100 m²) in heavy soils.

Historic mouldboard plough development

The basic plough, with coulter, ploughshare and mouldboard remained in use for a millennium. Major changes in design did not become common until the Age of Enlightenment, when there was rapid progress in design. Thomas Jefferson, the renowned statesman, was the first to bring theoretical knowledge to the design and the construction of the mouldboard. Writing in 1788, he referred to the curves which should characterize a mouldboard, and said: "The offices of the mouldboard are to receive the sod after the share has cut under it, to raise it gradually and to recover it. The fore end of it should, therefore, be horizontal, to enter the sod, and the hind end perpendicular, to throw it over; the intermediate surface changing gradually from the horizontal to perpendicular. It should be as wide as the furrow, and of a length suited to the construction of the plough."

While Jefferson succeeded very well in using the experimental ploughs which he made, the time was not yet ripe for the general adoption of his ideas, and his work was lost for a generation, until it was taken up and improved upon by Wood and later inventors.

The first mouldboard ploughs could only turn the soil over in one direction (conventionally always to the right), as dictated by the shape of the mouldboard, and so the field had to be ploughed in long strips, or *lands*. The mouldboard is covered with iron strips to slow wear on the mouldboard and prevent soil from sticking to it. It has an iron share connected to the drawbar (or beam) with an iron coulter (or cutter). The mouldboard plough actually flipped the soil over and turned a true furrow. The plough was usually worked clockwise around each land, ploughing the long sides and being dragged across the short sides without ploughing. The length of the strip was limited by the distance oxen (or later horses) could comfortably work without a rest, and their width by the distance the plough could conveniently be dragged.

These distances determined the traditional size of the strips: a furlong, (or "furrow's length", 220 yards (200 m)) by a chain (22 yards (20 m)) – an area of one acre (about 0.4 hectares); this is the origin of the acre. The one-sided action gradually moved soil from the sides to the centre line of the strip. If the strip was in the same place each year, the soil built up into a ridge, creating the ridge and furrow topography still seen in some ancient fields.

The mouldboard plough greatly reduced the amount of time needed to prepare a field, and as a consequence, allowed a farmer to work a larger area of land. In addition, the resulting pattern of low (under the mouldboard) and high (beside it) ridges in the soil forms water channels, allowing the soil to drain. In areas where snow buildup is an issue, this lets farmers plant the soil earlier, as the snow runoff drains away more quickly.

Figure 7-27 shows the basic parts of the modern plough, other parts not shown or labeled include the frog (or frame), runner, landside, shin, trash board, and stilts (handles).

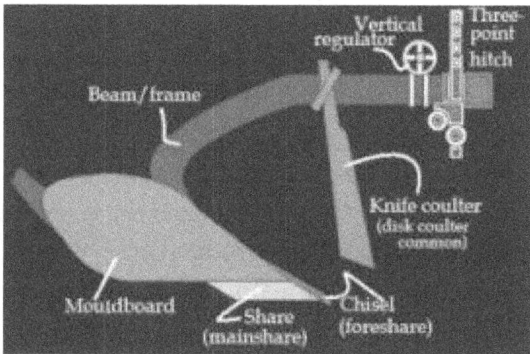

Figure 7-27: Schematic drawing of a contemporary plough

A *runner* extending from behind the share to the rear of the plough controls the direction of the plough, because it is held against the bottom land-side corner of the new furrow being formed. The holding force is the weight of the sod, as it is raised and rotated, on the curved surface of the mouldboard. Because of this runner, the mouldboard plough is harder to turn around than the scratch plough, and its introduction brought about a change in the shape of fields – from mostly square fields into longer rectangular "strips" (hence the introduction of the furlong).

A *coulter* (or skeith) could be added to cut vertically into the ground just ahead of the *share* (in front of the **frog**), a wedge-shaped cutting edge at the bottom front of the *mouldboard* with the landside of the frame supporting the undershare (below-ground component).

The upper parts of the frame carry (from the front) the coupling for the motive power (horses), the coulter and the landside frame. Depending on the size of the implement, and the number of furrows it is designed to plough at one time, a forecarriage with a wheel or wheels (known as a furrow wheel and support wheel) may be added to support the frame (wheeled plough). In the case of a single-furrow plough there is only one wheel at the front and handles at the rear for the ploughman to steer and manoeuvre it.

When dragged through a field the coulter cuts down into the soil and the share cuts horizontally from the previous furrow to the vertical cut. This releases a rectangular strip of sod that is then lifted by the share and carried by the mouldboard up and over, so that the strip of sod (slice of the topsoil) that is being cut lifts and rolls over as the plough moves forward, dropping back to the ground upside down into the furrow and onto the turned soil from the previous run down the field.

The ploughshare spread the cut horizontally below the surface, so when the mouldboard lifted it, a wider area of soil was turned over. Each gap in the ground where the soil has been lifted and moved across (usually to the right) is called a *furrow*. The sod that has been lifted from it rests at about a 45 degree angle in the next-door furrow and lies up the back of the sod from the previous run.

In the basic mouldboard plough the depth of the cut is adjusted by lifting against the runner in the furrow, which limits the weight of the plough to what the ploughman could easily lift. This limited the construction to a small amount of wood (although metal edges were possible). These ploughs were fairly fragile, and were not suitable for breaking up the heavier soils of northern Europe. The introduction of wheels to replace the runner allowed the weight of the plough to increase, and in turn allowed the use of a much larger mouldboard faced in metal.

Figure 7-28: Disc ploughs in use in Australia, circa 1900

Stump-jump plough: The Stump-jump plough was an Australian invention of the 1870s, designed to cope with the breaking up of new farm land that contains many tree stumps and rocks that would be very expensive to remove. The plough uses a moveable weight to hold the ploughshare in position. When a tree stump or other obstruction such as a rock is encountered, the ploughshare is thrown upwards, clear of the obstacle, to avoid breaking the plough's harness or linkage; ploughing can be continued when the weight is returned to the earth after the obstacle is passed.

A simpler system, developed later, uses a concave disc (or a pair of them) set at a large angle to the direction of progress, which uses the concave shape to hold the disc into the soil – unless something hard strikes the circumference of the disk, causing it to roll up and over the obstruction. As the arrangement is dragged forward, the sharp edge of the disc cuts the soil, and the concave surface of the rotating disc lifts and throws the soil to the side. It does not make as good a job as the mouldboard plough (but this is not considered a disadvantage, because it helps fight the wind erosion), but it does lift and break up the soil.

Type of ploughing method

Single-sided ploughing: The first mouldboard ploughs could only turn the soil over in one direction (conventionally always to the right), as dictated by the shape of the mouldboard, and so the field had to be ploughed in long strips, or *lands*.

Figure 7-29: Single-sided ploughing in a ploughing match

The plough was usually worked clockwise around each land, ploughing the long sides and being dragged across the short sides without ploughing. The length of the strip was limited by the distance oxen (or later horses) could comfortably work without a rest, and their width by the distance the plough could conveniently be dragged.

Turnwrest plough: The turnwrest plough allows ploughing to be done to either side. The mouldboard is removable, turning to the right for one furrow, and then being moved to the other side of the plough to turn to the left (the coulter and ploughshare are fixed). In this way adjacent furrows can be ploughed in opposite directions, allowing ploughing to proceed continuously along the field and thus avoiding the ridge and furrow topography.

Reversible plough: The reversible plough has two mouldboard ploughs mounted back-to-back, one turning to the right, the other to the left. While one is working the land, the other is carried upside-down in the air. At the end of each row, the paired ploughs are turned over, so the other can be used. This returns along the next furrow, again working the field in a consistent direction.

Figure 7-30: A four-furrow reversible Kverneland plough

Balance plough: The balance plough had two sets of ploughs facing each other, arranged so when one was in the ground, the other set was lifted into the air. When pulled in one direction, the trailing ploughs were lowered onto the ground by the tension on the cable. When the plough reached the edge of the field, the other engine pulled the opposite cable, and the plough tilted (balanced), putting the other set of shares into the ground, and the plough worked back across the field. One set of ploughs was right-handed, and the other left-handed, allowing continuous ploughing along the field, as with the turnwrest and reversible ploughs.

Figure 7-31: A German balance plough

In America the firm soil of the Plains allowed direct pulling with steam tractors, such as the big Case, Reeves or Sawyer-Massey breaking engines. Gang ploughs of up to fourteen bottoms were used. Often these big ploughs were used in regiments of engines, so that in a single field there might be ten steam tractors each drawing a plough. In this way hundreds of acres could be turned over in a day. Only steam engines had the power to draw the big units. When internal combustion engines appeared, they had neither the strength nor the ruggedness compared to the big steam tractors. Only by reducing the number of shares could the work be completed.

Figure 7-32: Northern Great Plains, 1880-1920

Loy (spade) ploughing: Loy ploughing was a form of manual ploughing in Ireland, on very small farms—or on very hilly ground, where horses could not work or where farmers could not afford them (Hughes, 2011). It was used up until the 1960s in poorer land (White, *Medieval Technology*). This suited the moist climate of Ireland as the trenches formed by turning in the sods providing drainage. It also allowed the growing of potatoes in bogs as well as on mountain slopes where no other cultivation could take place (Bell, 1996).

Modern ploughs

Modern ploughs are usually multiple reversible ploughs, mounted on a tractor via a three-point linkage. These commonly have between two and as many as seven mouldboards – and **semi-mounted** ploughs (the lifting of which is supplemented by a wheel about halfway along their length) can have as many as eighteen mouldboards. The hydraulic system of the tractor is used to lift and reverse the implement, as well as to adjust furrow width and depth.

Figure 7-33: Modern farm plough

The ploughman still has to set the draughting linkage from the tractor so that the plough is carried at the proper angle in the soil. This angle and depth can be controlled automatically by modern tractors. As a complement to the rear plough a two or three mouldboards-plough can be mounted on the front of the tractor if it is equipped with front three-point linkage. On modern ploughs and some older ploughs, the mouldboard is separate from the share and runner, so these parts can be replaced without replacing the mouldboard.

Specialist ploughs

Scottish hand plough: This is a variety of ridge plough notable in that the blade points towards the operator. It is used solely by human effort rather than with animal or machine assistance, and is pulled backwards by the operator, requiring great physical effort. It is particularly used for second breaking of ground, and for potato planting. It is found in Shetland, some western crofts and more rarely Central Scotland. The tool is typically found on small holdings too small or poor to merit use of animals.

Chisel plough: The *chisel plough* is a common tool to get deep tillage (prepared land) with limited soil disruption. The main function of this plough is to loosen and aerate the soils while leaving crop residue at the top of the soil. This plough can be used to reduce the effects of compaction and to help break up plough pan and hardpan.

Unlike many other ploughs the chisel will not invert or turn the soil. This characteristic has made it a useful addition to no-till and low-till farming practices that attempt to maximize the erosion-prevention benefits of keeping organic matter and farming residues present on the soil surface through the year. Because of these attributes, the use of a chisel plough is considered by some to be more sustainable than other types of plough, such as the mouldboard plough.

Figure 7-34: Animal drawn chisel plough (T. Friedrich)

The ploughing tines are at the rear; the refuse-cutting coulters at the front. The chisel plough is typically set to run up to a depth of eight to twelve inches (200 to 300 mm). However some models may run much deeper. Each of the individual ploughs, or shanks, are typically set from nine inches (229 mm) to twelve inches (305 mm) apart. Such a plough can encounter significant soil drag, consequently a tractor of sufficient power and good traction is required. When planning to plough with a chisel plough it is important to bear in mind that 10 to 15 horsepower (7 to 11 kW) per shank will be required.

Figure 7-35: A modern John Deere 8110 farm tractor using a chisel plough

Ridging plough: A ridging plough is used for crops, such as potatoes or scallions, which are grown buried in ridges of soil using a technique called ridging or hilling. A ridging plough has two mouldboards facing away from each other, cutting a deep furrow on each pass, with high ridges either side. The same plough may be used to split the ridges to harvest the crop.

Auto-plough: Early on in tractor design was the auto-plough. The Auto-plough was a tractor with the plough blades already mounted to the bottom of the tractor. The Figure to the Left is a 1912 Hackney auto plough. Notice wheel in front of the tractor. You would turn this wheel to start the engine. It is believed to be the only running one of its kind housed at the Martha and Dale Hawk Museum in North Dakota

Figure 7-36: The auto plough

Mole plough (widely called **subsoiler):** The *subsoiler* or *mole plough* allows under-drainage to be installed without trenches, or it breaks up deep impermeable soil layers that impede drainage. It is a very deep plough, with a torpedo-shaped or wedge-shaped tip, and a narrow blade connecting this to the body. When dragged through the ground, it leaves a channel deep under the ground, and this acts as a drain. Modern mole ploughs may also bury a flexible perforated plastic drain pipe as they go, making a more permanent drain – or they may be used to lay pipes for water supply or other purposes. Similar machines, so called pipe-and-cable-laying ploughs, are even used under the sea, for the laying of cables, as well as preparing the earth for side-scan sonar in a process used in oil exploration.

Paraplough: The paraplough or paraplough is a tool for loosening compacted soil layers 12 to 16 inches deep and still maintain high surface residue levels.

Spade plough: The spade plough is designed to cut the soil and turn it on its side, minimizing the damage to the earthworms, soil microorganism, and fungi. This helps maximize the sustainability and long term fertility of the soils. In 1935, Agronomists Frank Duley and Jouette Russel conducted the first research on conservation tillage using a special spade plough called sweep plough. Between 1950 and 60s, Eugene McKibben conducted theoretical and applied researches in the soil dynamics of ploughs and other tillage equipment, and directs the USDA research programs in mechanization.

Effects of mouldboard ploughing

Ploughing leaves very little crop residue on the surface, which otherwise could reduce both wind and water erosion. Over-ploughing can lead to the formation of hardpan. Typically farmers break up hardpan up with a subsoiler, which acts as a long, sharp knife to slice through the hardened layer of soil deep below the surface. Soil erosion due to improper land and plough utilization is possible. Contour ploughing mitigates soil erosion by ploughing across a slope, along elevation lines.

Figure 7-37: A 10-20 HP Mogul tractor breaking sod with a 4-bottom plough

Alternatives to ploughing, such as the no till method, have the potential to actually build soil levels and humus. These may be suitable to smaller, more intensively cultivated plots, and to farming on poor, shallow or degraded soils that ploughing would further.

Harrowing

Secondary to the ploughing operations in soil preparation before planting, is the harrow operation. A typical harrow consists of a wooden or metal framework bearing metal disks, teeth, or sharp projecting points, called tines, which is dragged over ploughed land to pulverize the clods of earth and level the soil. Harrows smoothened out rough, clumpy soil to ensure an even planting. Harrows are also used to uproot weeds, aerate the soil, and cover seeds.

History and development of harrow

The harrow's historic significance lies in the Chinese relationship to the themes of the Gold Rushes experience, market gardens, racial antagonism and fear of the exotic and the future unknown. The harrow provides a research tool for historians to explore the culture of the Chinese on the market garden in regional New South Wales. The Chinese gardeners made use of harrow to cultivate the surface of the soil in the Wagga Wagga area of the Riverina and western New South Wales, Australia in the 19th century.

Figure 7-38: Chinese harrow, c.1860-70s

The harrow looked like a large rake with rows of teeth and having aesthetic significance relating to the design and manufacture of 19th century improvised agricultural tools otherwise known as 'bush tools'.

Primitive harrows were made of twiggy branches drawn over the soil to smooth it; in India a ladder-like device of bamboo is still used. In modern large-scale farming, harrows are of varied types. Some are simply dragged behind a tractor or draft animal; some are suspended on wheels; many have levers to adjust the depth of the cut. There may be one or more gangs (sets) of cutting parts per harrow, and one or more harrows may be drawn at a time. In general, the harrow is similar to the cultivator, except that it penetrates the soil to a lesser depth.

The early harrows were square shaped with spikes attached to a wooden frame pulled by a horse or ox. Later the design was changed to bifurcate or triangular shape, which made it easier to pull though the soil by the horse or ox. The square harrow was used to even out fields already free of obstacles (such as tree stumps) and could cover more ground than a bifurcate harrow.

Figure 7-39: Clod harrow

The initial harrows were constructed out of simple wood frames with perhaps wooden stakes inserted as tines to help break up or smooth the ploughed ground to prepare a seedbed.

Figure 7-40: Early rectangular harrow (c.1850)

The double frame "Scotch" wooden harrow has iron spikes protruding from the wooden frame to till and smooth the soil. A bifurcate harrow was more easily used in field that had obstacles because it of its relative small size, sturdier and easier maneuverability.

Figure 7-41: Finlayson's harrow

Harrows showed the same progression of wood construction, some cast iron parts, nearly all iron to wholly steel that were initially pulled mostly by oxen, then horses, which were in turn replaced by tractors starting about 1900.

Figure 7-42: Clydesdales horse pulling a harrow

Harrows always consist of a rigid frame to which are attached steel teeth (tines), cupped steel discs, linked chains or other means of smoothing or cultivating the soil. Tine and chain harrows are often only supported by a rigid towing-bar at the front of the set.

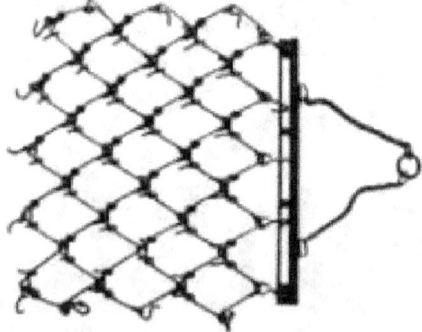

Figure 7-43: Chain harrow (Concord harrow)

Spike-tooth harrow: Spike-tooth harrows have rigid teeth, and spring-tooth harrows have curved tines that adjust to obstacles. By the turn of the 19th century, the iron spiked harrow had been replaced by the disk harrow, which is still in use today.

Figure 7-44: Spike harrow (Concord harrow)

Disk harrow: Disk harrow, which is next to the plough are the most widely used tillage implements, the saucer-shaped disks are set at angles to the line of pull for maximum pulverization. Some disc harrows, are made with revolving self-cleaning circular cupped steel discs that cut up, loosen, over turn and smooth the soil. On some soils larger versions of these discs can sometimes be used instead of ploughs.

Figure 7-45: Disc harrow (Concord harrow)

Offset disk harrow was introduced in USA in 1924. Other harrow types that were invented include drag tooth harrows, chain harrows and power harrows or cultivators, which have petrol engine, powered rotating L-shaped tines.

Rotary harrow: The rotary cross-harrow has power-driven rotating toothed disks; another type of harrow slices through topsoil and vegetation with curved knives. After tractors were introduced the farmer sat on the tractor while towing the harrow. Harrows seldom had riding accommodations provided for the teamsters.

History and development of cultivator

After the fields were ploughed, and the big clumps of soil broken up, the seed bed had to be prepared. The better the soil was broken up, the better the crops would grow in the soil. Early seed beds were prepared by hand with sticks, spades, or rakes before the invention of cultivators around the eighteenth century. A cultivator is a farm implement generally considered as being used for secondary tillage, they are often used for primary tillage in lighter soils instead of ploughing.

Figure 7-46: Early cultivators

One sense of the name refers to frames with teeth (also called shanks) that pierce the soil as they are dragged through it linearly. Another sense refers to machines that use rotary motion of disks or teeth to accomplish a similar result. Around the 18th century cultivators began to take over from the harrows to work the soil. Cultivators were originally drawn by draft animals (such as horses, mules, or oxen) or were pushed or drawn by people.

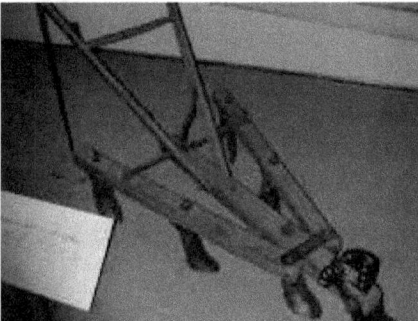

Figure 7-47: Triangular cultivator with iron frame

In about 1820 farmers were using single-row horse-drawn equipment. Henry Burden (1791-1871) in 1820 produced the first horse-drawn cultivator in USA and also invented a horseshoe machine. Horse-drawn cultivator was introduced to English farming by Jethro Tull, which was patented in 1733. In 1853, horse-drawn digger was patented by Robert Romaine, of Peterborough, Canada. In many American regions, shovel ploughs remained the common cultivating tools until the 1850's. By 1857 it was produced as Romaine-Crosskill digger with 14 hp steam engine. Cultivating machine consisting of a huge corrugated drum that crushed the clods, followed by a harrow, was introduced by Messrs. Blackburn, Derby, UK in 1857.

Figure 7-48: Horse drawn cultivator

By the late 1860's, horse-drawn two-row cultivators with a seat for the farmer, also called a sulky, were being used in the Midwest. Cultivators were then mounted on wheels which gave them a great advantage to the early models. These machines could control the depths at which they went into the soil. In 1878, Ohio Cultivator Co. was founded by Harlow Case Stahl (1850-1941), who is credited with perfecting the riding cultivator. By 1937, Noble Blade cultivator invented by Canadian Charles S. Noble, Nobleford, Alberta, Canada was patented for prairie lands. It sheared the stubble below the soil surface and reduced erosion.

Figure 7-49: Wheel mounted cultivator

The powered rotary hoe was invented by Arthur Clifford Howard who, in 1912, began experimenting with rotary tillage on his father's farm at Gilgandra, New South Wales, Australia. Initially using his father's steam tractor engine as a power source, he found that ground could be mechanically tilled without soil-packing occurring, as was the case with normal ploughing. His earliest designs threw the tilled soil sideways, until he improved his invention by designing an L-shaped blade mounted on widely spaced flanges fixed to a small-diameter rotor.

Figure 7-50: Howard rotary hoe

Meanwhile, in North America during the 1910s, tractors were evolving away from traction engine-sized monsters toward smaller, lighter, more affordable machines. The Fordson tractor especially had made tractors affordable and practical for small and medium family farms for the first time in history. Cultivating was somewhat of an afterthought in the Fordson's design, which reflected the fact that even just bringing practical motorized tractive power alone to this market segment was in itself a milestone. This left an opportunity for others to pursue better motorized cultivation.

Between 1915 and 1920, various inventors and farm implement companies experimented with a class of machine referred to as motor cultivators, which were simply modified horse-drawn shank-type cultivators with motors added for self-propulsion. This class of machines found limited market success. But by 1921 International Harvester had combined motorized cultivating with the other tasks of tractors (tractive power and belt work) to create the Farmall, the general-purpose tractor tailored to cultivating that basically invented the category of row-crop tractors.

Figure 7-51: Farmall C-254 cultivator, 1949

Description and uses

Cultivators are often similar in form to chisel ploughs, but their goals are different. Cultivator teeth work near the surface, usually for weed control, whereas chisel plough shanks work deep beneath the surface, breaking up hardpan. Consequently, cultivating also takes much less power per shank than does chisel ploughing.

Figure 7-52: Tines of a rotary cultivator

Cultivators are usually either self-propelled or drawn as an attachment behind either a two-wheel tractor or four-wheel tractor. For two-wheel tractors they are usually rigidly fixed and powered via couplings to the tractors' transmission.

Figure 7-53: Cultivator

For four-wheel tractors they are usually attached by means of a three-point hitch and driven by a power take-off (PTO) system. Drawbar hookup is also still commonly used worldwide. Draft-animal power is sometimes still used today, being somewhat common in developing nations although rare in more industrialized economies. The cultivator may be an implement trailed after the tractor via a drawbar; mounted on the three-point hitch; or mounted on a frame beneath the tractor.

Uses: Cultivators stir and pulverize the soil, either before planting (to aerate the soil and prepare a smooth, loose seedbed) or after the crop has begun growing (to kill weeds-controlled disturbance of the topsoil close to the crop plants kills the surrounding weeds by uprooting them, burying their leaves to disrupt their photosynthesis, or a combination of both). Small toothed cultivators pushed or pulled by a single person are used as garden tools for small-scale gardening, such as for the household's own use or for small market gardens.

Types of cultivators

Row crop cultivators: Row crop cultivators, sometimes referred to as *sweep cultivators*, are usually raised and lowered by a three-point hitch and the depth is controlled by gauge wheels. *Sweep cultivators* commonly have two center blades that cut weeds from the roots near the base of the crop and turn over soil, while two rear sweeps further outward than the center blades deal with the center of the row, and can be anywhere from 1 to 5 rows wide.

Figure 7-54: Homemade sweep cultivator (Notice the inner and outer "sweep" blades)

Garden cultivators: Small tilling equipment, used in small gardens such as household gardens and small commercial gardens, can provide both primary and secondary tillage. For example, a rotary tiller does both the "ploughing" and the "harrowing", preparing a smooth, loose seedbed. It does not provide the row-wise weed control that cultivator teeth would. For that task, there are single-person-pushable toothed cultivators.

Cultivator trademarks

A small rotary hoe for domestic gardens was known by the trademark Rototiller and another, made by the Howard Group, who produced a range of rotary tillers, was known as the Rotavator.

Rototiller: The small rototiller is typically propelled forward via a (1–5 horsepower or 0.8–3.5 kilowatts) petrol engine rotating the tines, and do not have powered wheels, though they may have small transport/level control wheel(s). The slower a rototiller moves forward, the more soil tilth can be obtained. Rototilling is much faster than manual tilling, but notoriously difficult to handle and exhausting work, especially in the heavier and higher horsepower models.

Figure 7-55: A compact rototiller

Rotavator: The trademarked word "Rotavator" is one of the longest single-word palindromes in the English language. Unlike the rototiller, the self-propelled Howard rotavator (Figure 7-56) is equipped with a gearbox and driven forward, or held back, by its wheels. The gearbox enables the forward speed to be adjusted while the rotational speed of the tines remains constant which enables the operator to easily regulate the extent to which soil is engaged.

Figure 7-56: A Howard rotavator attached to the tractor 3-point linkage

For a two-wheel tractor rotavator this greatly reduces the workload of the operator as compared to a rototiller. These rotavators are generally heavy duty, come in higher power (4-18 horsepower or 3-13 kilowatts) with either petrol or diesel engines and can cover much more area per hour.

Mini tiller: Mini tillers are a new type of small agricultural tillers or cultivators used by farmers or homeowners. These are also known as power tillers or garden tillers. Compact, powerful and, most importantly, inexpensive, these agricultural rotary tillers are providing alternatives to four-wheel tractors and in the small farmers' fields in developing countries are more economical than four-wheel tractors.

Figure 7-57: An F210 Honda mini tiller

7.3 Historic developments in crop planting and seed sowing

Early seed sowing methods

Br Until the 18th century when dibbers or dibbles were first used in numbers, the common method of sowing seed in the primitive agricultural era was by hand distribution or broadcast.

Seed broadcasting

Broadcast seeding probably originated in the valley of the Nile, where, after the water had subsided, a farmer could sow his seed and drive sheep over the ground or go over it with a brush harrow or plough. The parable of the seed sower as told by Jesus Christ in the epistle of Luke Chapter 8 Verses 5-8, inferred that early planting system practiced in the Old Stone Age was done by hand. The seeds were thrown into the air or broadcast to scatter the seed randomly on the field. This system made it more difficult to weed and harvest the crop.

The following methods of seed broadcast were practiced in the primitive and medieval times.

Aprons – seedlips broadcasting

This system has the sower walked behind the plough carrying the seed in an apron or a container called a Seedlip. The apron or sowing sheet was nothing more than a linen cloth wrapped around the shoulder and an arm or attached to the waist. Seedlips generally contained between four and six gallons of seed corn (1/2 to ¾ of a bushel). They were hung on the sower's left side by neck or should straps and supported by whichever hand was not sowing.

Seed fiddles broadcaster

Portable, hand-operated broadcast-sow machines were introduced from America from 1850 onwards. The most novel, if not the most popular, of these was the seed fiddle which took its name from the fiddle-like action required to distribute the seed. The device was suspended to one side of the operator by a shoulder strap and supported by one arm.

Figure 7-58: Seed fiddle, courtesy of Roy Brigden

It consisted of canvas seed bag housed in a small rectangular box frame with an horizontally mounted finned disc which was rotated in alternating revolutions by a leather thronged bow. As the bow was moved from one side to the other with each step the sower took, a regulated amount of seed dropped from the bag on to the revolving disc which scattered it in a wide arc from 16 to 24 feet across according to the type of seed.

Uses: Used mainly for grass and clover many of them were imported but some, notably the 'Aero' brand were made in Kilmarnock and distributed all over the country until recent times. In 1940 they sold for 27 shillings and sixpence plus a shilling for carriage!

Cyclone seed sower

Other manually operated broadcast seeders from America appeared at roughly the same time as the seed fiddle. Instead of a bow the horizontal seed distributing disc of the cyclone seed sower was briskly turned by hand crank and gears spreading wheat or rye seed over an area of 30 feet or more. The seed bag rested on a simple platform which also supported the machinery. The Chicago seeder had a similar mechanism but was also supplied with a fiddle bow.

Cahoon seed sower

Cahoon's broadcast seed sower patented in the USA in 1861 has a side mounted hand crank with high ratio gearing rotated an open ended distributor drum mounted at the front. The drum contained four radial fins or ribs which scattered the seed in all directions as it fell from the sack through a graduated slide opening. Another form had two drums paced one above the other. The upper revolving in one direction received half the seed, the lower turning in the opposite direction received the other half of seed which possibly resulted in a more even distribution.

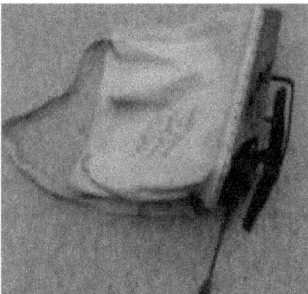

Figure 7-59: Cahoon's broadcast seeder

Seed box broadcasters

Broadcast seed boxes were already in use in the USA during the 1830's. This unusual 19th century device was used for sowing clover, turnip and other small seeds. The Chinese have a kind of wheelbarrow seeder with hollow teeth which draws furrows and drops the seed, and it is claimed that this implement has been used for ages.

Figure 7-60: Broadcast barrow, courtesy of Roy Brigden

It comprised of a long narrow box fitted with a single or divided hinged lid which measured 2770 mm (9 feet) in length, 66 mm wide and 60 mm in depth. The interior was divided equally into twelve separate compartments each perforated with six equidistant holes in the bottom. These were overlaid with a piece of tin plate drilled with a corresponding number of holes of smaller diameter which allowed a single seed to pass through.

Figure 7-61: English wheelbarrow seeder, 1820

To regulate the amount of seed falling from every compartment each hole could be closed off by a pivotal tin shutter shaped like a finger fixed nearby. By moving a perforated wooden slide fitted underneath the box the seed supply could be shut off. Some contained a zinc plated base with only a single aperture per compartment while other examples were fitted with copper slides along the bottom.

The box hung from a shoulder strap and was jerked from side to side with each step the sower took. Each sideways movement delivered a single seed from each opening. By varying the length of stride the amount of seed discharged in any given area could be regulated. Much however depended on the attentiveness of the operator; who was also required to constantly replenish each compartment.

Soon it was realized that no-till sowing technology requires seeders that will effectively penetrate untilled soil and place the seed at the optimum depth for rapid plant emergence led to the later development of dibbers.

Dibber

A dibber or dibble is a pointed wooden stick for making holes in the ground so that seeds, seedlings or small bulbs can be planted. The dibber was first recorded in Roman times and has remained mostly unchanged since. The earliest handcrafted examples of dibbers were made from a piece of horn, bone or a simple branch carved into a point and hardened by the heat of a fire.

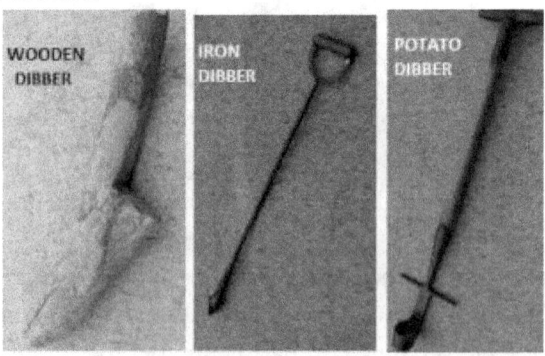

Figure 7-62: Dibbers

Typically, one farmhand would walk ahead with the dibber making holes, while a second would plant seeds and fill the holes. In Roman times farmers used handled dibbers of metal & wood to plant their crops.

In the 18th and 19th centuries, farmers would use long-handled dibbers of metal or wood to plant crops. One man would walk with a dibber making holes while a second man would plant seeds in each hole and fill it in. Dibbers come in a variety of designs including the straight dibber, T-handled dibber, trowel dibber, and L-shaped dibber. In some countries (such as New Zealand) the term is also often used to refer to a mattock. Dibber technology had improved over the centuries and it was not until the periods of Renaissance that dibbers became a manufactured item, some made of iron for penetrating harder soils and clay.

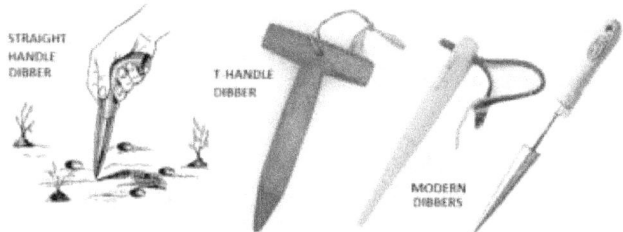

Figure 7-63: Dibbers

The modern day seed dibber designs have markings graduated at every 1cm to make it easy to sow seeds to the right depth so that more of them germinate. Improved hand-held dibbers are still in use on small scale farms, even well into the 21st century.

Figure 6-64: Hand-held dibbler in use

Dibber was very effective and still proves valuable alongside today's highly technical tools, but also very tedious and time consuming, and then the idea for dropping seeds through a tube appeared first in Mesopotamia about 1500 B. C. which brought about the development of seed drills.

Seed drills

Seed drill is a kind of sowing device that accurately plant seeds of different crops in the soil and covers them thereafter. The invention of seed drill was revolutionary to agricultural field because the method used previously in planting seed (broadcasting) was very inefficient and wasteful. Less people were needed to seed the land, so there was a decrease in needs for slaves. Also, because more crops were grown, more people were able to get their hands on foods. Seed drill was very popular at that times, so many people started a business that sold seed drills with many other features added to it, even nowadays. For example, there were variable types of hoppers and tubes for different purposes.

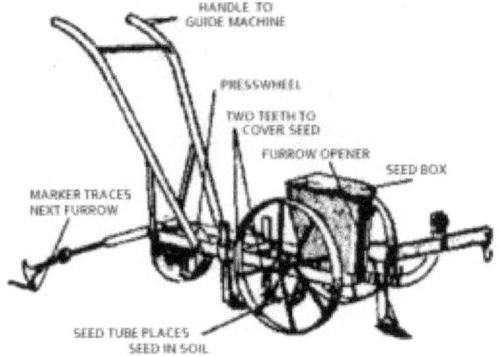

Figure 7-65: Typical seed drill

Seed drill is formed from a bunch of seeds loaded in numbers of tubes that allows seeds that are planted to keep certain distances from each other to allow better growth of seeds planted. In real life uses, a seed drill is rolled front to make a certain knife to dig the soil. And from a metering mechanism on hoppers (hoppers contain bunch of seeds), it makes certain number of seeds to fall into tubes in certain time, and those seeds go through the tube and falls into holes that seed drill made previously. The holes can be covered with built-in rake.

The invention of seed drill had impacted our modern society greatly because nowadays there are more people who get enough amount of food, for reasons of many agricultural machines developed, seed drills starting with seed drills. Modern society is better off than the pre-industrial society as a result of seed drill invention.

Seed drills development

Before seed drill was invented, farmers had to plant seed by their own hands. But not only was it very tiring, it was also inaccurate and the seeds were not distributed equally, which led to poor production speed. The invention of seed drill improved speed of seeding and better accuracy than it was originally. The first trace of a seeding machine found in history was an Assyrian drill used many centuries before the birth of Christ, a reproduction of it being found on the Aberdeen "black stone," of the time of Esarhaddon, 680 B. C. The implement had a mouldboard made from a round stick of toughened wood, with a tongue and handles attached. In the rear of the plough point was attached a bowl-shaped hopper, supported upon a hollow standard, through which seed passed to the furrow, and was covered by the turned furrow falling back upon it.

There are records that Sumerians and Ancient Chinese people use the seed drill. While the Babylonians used the primitive single-tube seed drills around 1500 BC, the invention never reached Europe. The Chinese invented a multi-tube iron seed drills in the 2nd century BC. There were no documentary evidences on claims that seed drill was first introduced in Europe by the Chinese people.

Figure 7-66: Ox-drawn Chinese seed drill, 1637

The first known European seed drill developed was patented by a Venetian (Italian) named Camillo Torello in 1556. At that time, the continuous development of conservation tillage technologies led to active studies on the performance of seeders.

It is said that in Italy about the year 1600 A, D., a seeder running on two wheels and supporting a seed-box on its axle, was used. It was mounted on two wheels, the axle passing through the seed-box, on the bottom of which was a series of holes opening into an equal number of metal tubes or funnels, through which the seed was conducted to the ground. The fronts of the tubes, at their lower ends, were shaped somewhat like ploughshares, and were designed to make small furrows into which the seed dropped.

Several efforts were made during the sixteenth century by English inventors to perfect a seeding machine, and their machines may have worked well in the hands of the inventors, but were soon lost sight of and forgotten.

One machine by an unknown English inventor was manufactured and patented about 1664, and in 1669 John Evelyn presented one invention to the Philosophical Society of London, and there was a claim that an agent was appointed in London for its sale. The machine was attached to the "stilts" of a plough, behind, and consisted of a seed-box having a cylinder furnished with wheels to distribute the seed, which was dropped regularly in the furrow.

The success of direct drilling oilseed rape at Clayton and Frickley Farms near Doncaster encouraged manager Phil Barlow to look at similar methods for establishing cereals and beans. He observed that none of the land has been ploughed for the past seven years, and rape is established with a Sumo Trio and seed box, Mr Barlow and farm owner Charlie Warde-Aldam started to look round for suitable drills. They had heard of the Claydon drill and went to see it at LAMMA. They hired one to use at their farm and soon develop the Barlow's seed drill.

Figure 7-67: Barlow's bean drills

The greatest contribution to the early development of grain drills was made by Jethro Tull (1674-1741) in the eighteenth century. In a work which he published later in 1731, entitled, "Horse-hoeing Husbandry," he argued that grain and seed should not be sown by broadcast, but should be planted in rows or drills so as to admit of hoeing by horse power with proper implements. In 1701 Tull invented the first practical seed drill machine, consisting of a cylinder and regularly spaced holes that caught the seeds from the hopper above, cut small channels into the soil and dropped the seed into the channel (special furrow) below.

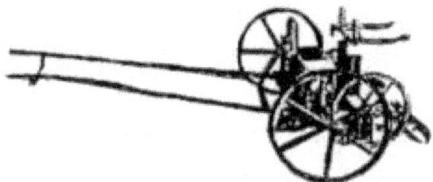

Figure 7-68: Jethro Tull's first practical seed drill

The machine also consisted of two seed-boxes with a coulter attached to each, and following each other; behind them followed a harrow to cover in the seed. His reason of having two separate deposits of seed, and at different depths, was that they might not sprout at the same time, and so perhaps escape the ravages of the fly. Jethro Tull invented other seed drills which could be pulled behind a horse. It consisted of a wheeled vehicle containing a box filled with grain. There was a wheel-driven ratchet that sprayed the seed out evenly as the seed drill was pulled across the field.

Figure 7-69: Jethro Tull improved drill

Jethro Tull's invention was met with skepticism and not really appreciated or accepted till after his death in 1741 (Blandford, 1976).

An English clergyman named James Cooke made many improvements in grain drill technology, some of which became a part of all British grain drills constructed since. He got a patent for spoon-feed drills with dropping cups and coulters in England by James Cooke in 1782. Barely a decade later, had he obtained patent for corn and small seeds seeding machine in 1799.

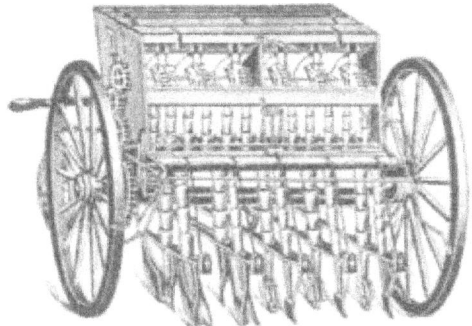

Figure 7-70: Cooke's grain drill

The manufacture of grain drills began at about 1840. A few drills had been brought over from England and introduced, and efforts had been made to establish the manufacture of those machines, but nothing permanent resulted. The first patent on a seeding machine in America was granted in 1799, and up to 1836, when the Patent Office records were burnt, patents had been granted to about thirty inventors in this line. It does not seem, however, that anything valuable had been contributed to the art beyond what we have noticed about the English inventors. The most important inventions that were left to be discovered were in the feed and in adjusting devices that today distinguish American drills.

By 1834, a black inventor named Henry Blair (1804-1860) patented a corn planter in USA, and in 1836, Blair also patented a cotton seed planter. The first important patent record was granted in 1835, and re-issued in 1838. It was on a machine designed to sow lime and plaster, and as re-issued showed that the invention was intended to sow grain, also. In 1837 another patent was granted, covering the application of centrifugal force, to sow lime, plaster and small grain. In 1838 a patent was issued for a grain drill in which a spring arm attached to a horizontal shaft revolved within the hopper and agitated the grain over the mouths of the tubes through which it was distributed.

Deere & Company was founded by John Deere (1804-1886), began manufacturing of planting equipment in 1839. One of the next innovations was a two row seed drill. Two-row horse-drawn corn planter was patented by D. S. Rockwell in 1839.

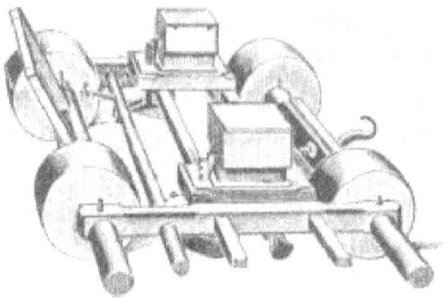

Figure 7-71: Rockwell's two-row drill, 1839

At about 1840, several efforts had been made in the USA to establish the manufacture of the machines, but nothing permanent resulted. However, on August 25, 1840, J. Gibbons, of Adrian, Mich., patented a grain drill with cavities to deliver seed, and a device for regulating its volume; and in 1841 he also patented a distributing cylinder, having several rows of cavities around its periphery, in combination with a hopper. These four patents were the only ones issued in six years, two of them; it will be noticed, being on broadcast seeding devices, and two on drills.

A simple single row seed drill made by E. C. Fairchild (Figure 7-62) claimed it could plant 6-10 acres a day!

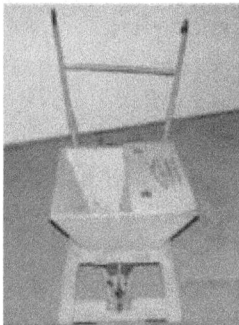

Figure 7-72: Fairchild's single row drill

Among the early inventors who made substantial improvements in the invention of drills, were Moses Pennock and Samuel Pennock (M. & S. Pennock), of East Marlboro, Pa., who made considerable progress in the development of 'cylinder drills'. Their first patent, dated March 12, 1841, and was re-issued on Oct. 30, 1849. It covered the simultaneous throwing into and out of operation by a lever of each seeding cylinder, and its corresponding tube and drill, and made so as to use any number of hoes desired. It covered also an arrangement of spur wheels for connecting the seed cylinders and hoppers to the shaft, so that they could be thrown into and out of gear when the drill is in motion.

Pennock drill consists of a seed box (hopper), slider arm, rollers and spur gears mounted on a rectangular wooden frame and equipped with adjustable lever arm for one or two horse attachment (Figure 7-73). Many other patents were issued to this firm, most of them covering improvements in cylinder drills, in which a series of cylinders operated over a series of hoes or tubes. Richard Jordan Gatling (1818-1903) made contributions to the development of cotton seed sowing drills around 1844. This was not automatic so the field would have to be marked and then the seeds released by the pulling of a lever.

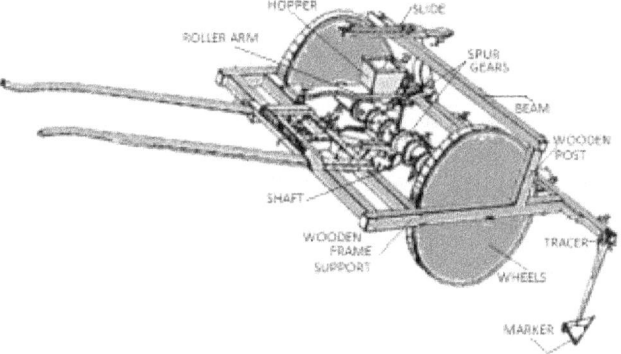

Figure 7-73: Samuel & Pennock lever drill, 1841

Around 1850 many different styles of mechanical planters appeared, but a number of reasons prevented the tools from achieving general acceptance. Primarily, farmers did not believe enough time was saved by the planter to justify purchasing one. To use the planter, a farmer would fill the wooden canister with seed and then pull up on the large handle to release one.

In some regions seed corn were tarred before it was planted to prevent birds from eating them. However, tarred corn cannot be put through this type of planter as it would gum up the mechanism. Another reason for not using these planters was the belief that corn grows better out of hill than a furrow. In order to plant quickly using a planter, furrows must be dug, hills cannot be used. Not until the turn of the century, after many improvements were made, were mechanical planters considered a standard farm implement.

In the years following 1850, patents were issued on grain drills at frequent intervals, and it is unnecessary for us to follow them in detail. By this time three different classes of drills were in the field, distinguished by their feeding devices. The first, of cylinder drills as built by the Pennocks and others; the second, as slide drills, in which the distribution was effected or governed by means of a slide; the third class, the force feed drills, which were then coming into use.

The first patent on a force feed for a grain drill was issued Nov. 4, 1851, to N. Foster, G. Jessup, H. L. and C. P. Brown, this invention introducing the name "force feed."

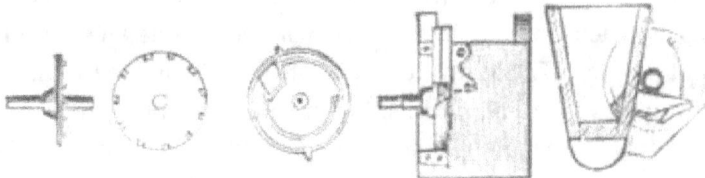

Figure 7-74: Forster, Jessup & Brown's force feed drill

These parties had been associated in the manufacture of grain drills at Palmyra, N. Y., since 1849.

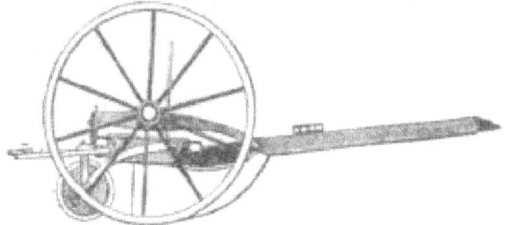

Figure 7-75: Side view of Geo. W Brown's first planter, 1853

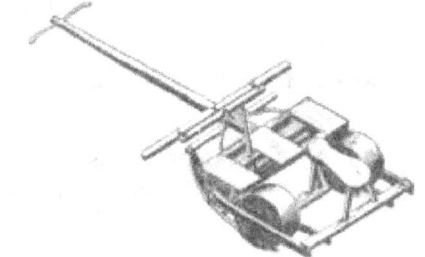

Figure 7-76: Geo. W Brown's planter, 1855

The first invention of a planter to drop in check automatically was accredited to M. Robbins, of Cincinnati, Ohio, whose patent (Figure 7-77) of Feb. 10, 1857, covered a reversible hopper, an arm with vibrating claw or tappet connected with the seeding mechanism, in combination with a jointed rod or chain provided with buttons.

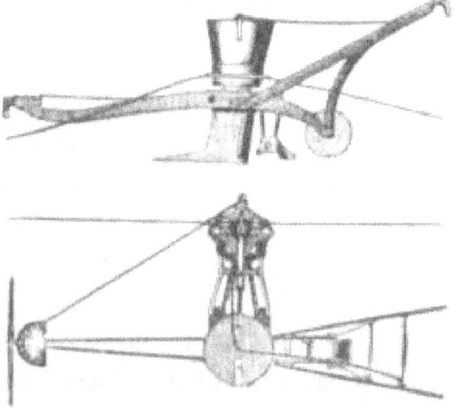

Figure 7-77: Robbins planter, 1857

This patent was re-issued Feb. 9, 1358, with three claims, the first covering the dropping of seed from a plough or drill by means of an anchored chain or its equivalent, the second claim covering the chain or cord, and the third claiming an arm with a vibrating claw or tappet, or equivalent devices, operating the seed discharging mechanism.

Mr. Robbins' invention was practically a one horse drill, with the chain or rod attached as patented, and it did not become known as a "check-rower." This name was given to later inventions of the Haworths and others, who had in view a separate attachment to be put on any planter. A few planters Robbins made worked well, but they were not practical and he died poor.

The next patent following that of Robbins was issued to John Thompson and John Ramsay, of Aledo, Ill., Sept. 29, 1864, and covered "the employment or use of a wire or cord, provided with knots at a suitable distance apart, and applied to the machine substantially as shown in connection with anchors. April 24, 1866, W. W. Hubbard, of Edinburg, Ind., obtained a patent for various improvements, which also was re-issued to the Haworths, March 27, 1877. This covered horizontal traversing bars for automatically moving check-row cords at the end of the field, and a movable arm for supporting on machine when turning. G. D. Haworth took out a patent on check-rower devices Feb. 22, 1870, and several others were issued later to the Haworths in the same line.

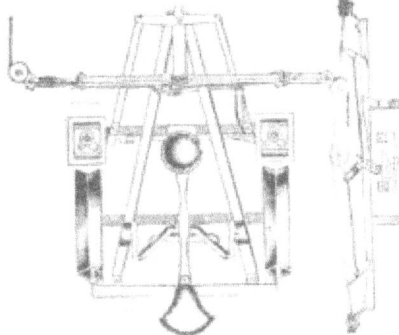

Figure 7-78: Haworth's check-row planter, 1867

Improvements now became necessary in the dropping mechanism. The slide drop had become too slow for use on the check-rower, and a rotary drop was therefore devised and adopted. With this invention the development of the more important features of the check-rower was completed, although many changes and improvements have been made from time to time by leading manufacturers in the details of its construction. In 1866 C. P. Brown patented a modification of the original Foster, Jessup & Brown feed, which has since been used in the Empire drill, and is known technically as the "single distributor."

About this time C. E. Patric, who had been in the employ of the Browns, moved to Macedon, N. Y., and he and Lyman Bickford took out several patents in 1867, covering the "double distributor." The distinguishing feature of this invention was a seed-wheel or disk with carrying flanges on each side, one chamber feeding coarse, bulky seeds, like oats, and the other being smaller, to sow wheat, rye, etc. The invention was adopted by Bickford & Huffman, of Macedon.

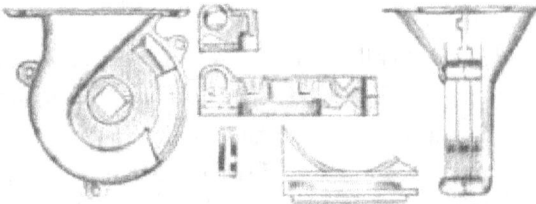

Figure 7-79: Patric & Bickford force feed planter, 1867

October 6, 1868, C. O. Gardiner, of Springfield, O., assignor to Thomas & Mast, secured a patent on a force feed that, with later improvements from the same inventor, became known as the Buckeye. Oct. 30, 1877, J. P. Fulghum patented a force feed principle that has been adopted by a number of prominent western manufacturers. Many other feeding devices have since been invented and introduced, but those we have noticed laid the foundation.

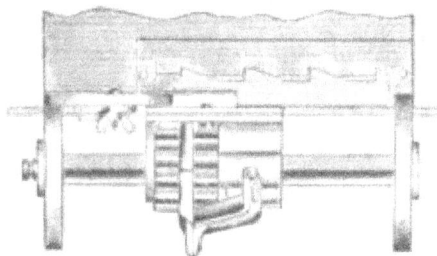

Figure 7-80: Fulghum force feed planter, 1877

Patents were granted at an early day on "adjustable, rank" drills, or those having devices for shifting the hoes from a straight rank to a staggered position. One of the most important was that of Charles F. Davis, Feb. 18, 1868

Improvements made in mechanical planters

Notable among several improvements made on mechanical planters within the century include

- Force-feed grain drill seeding implement patented in 1851 and developed to later become a part of products of International Harvester Co.
- Grain drill with depth control device invented by American Hiram Moore (1801-1874) in 1860.
- In 1855, check-row planter for corn and beans invented by Samuel Johnston (1835-1911), of Brockport, NY,

- The popular check-row corn planter patented by John Thompson and John Ramsey of Illinois in 1864, and
- The problem of automatic field marker solved by the development of automatic check row device for planting corn, usually in hills, developed using knotted cord (later wire) for tripping dropping of seed in 1875.

About a century later, in 1971, forced-air corn planters were introduced by International Harvester Co., preceded by electronic planter monitors introduced in the 1960s, most of which were manufactured by Dickey-john Corp., 1974. Deere & Co. introduced the Max-Emerge Planter, a plateless planter, with fingerpick- up seed metering. Below is a multiple row seed drill that could be adjusted to the amount of seeds and at what intervals they were released into the soil (Courtesy of Smithsonian American History Museum).

Figure 7-81: Multiple row seed drill

The modern day version of the seed drill is much larger and appears much more complicated. It uses air pressure to move the seeds through the tubes and into the soil. Yet, the basic principle is not much different from the original seed drill made by Tull.

Figure 7-82: Modern seed drill

Historic planting of stem cuttings demand a large quantity of human labour, time-consuming, unpleasant, and an arduous job. Therefore, stem planters suitable for use in local farms were developed to plant stakes such as cassava, sugarcane etc either on flat beds or in ridges and also to apply fertilizer at the same time. The machine consists of the main frame, the cutting unit, the planting unit, the fertilizer unit, the ridger, and the soil levelers all powered by tractor as the power source.

Figure 7-83: Early stem planting with machine

Mechanical thinning operation

The development of several mechanical thinners used a variety of technologies to sense plant material and selectively thin vegetable crops. Sensing devices included electric photo eyes, photocells that recognize plant colour and electric conducting circuits. These mechanical thinners, however, could not compete economically with hand thinning and were not adopted widely. Today there are a number of researchers and research institutions developing advanced computer technologies that can be used in conjunction with mechanical thinning and weeding.

Figure 7-84: Row crop thinner

7.4 Historic development in crop harvest and processing machinery

Traditional crop harvest and processing tools

Reaping tools

Reaping in agriculture implies the cutting of grain after maturity (when the grain had ripened sufficiently in the field and ready for harvest). The farmer used either a sickle, scythe, or cradle scythe to cut the crop in a process called reaping.

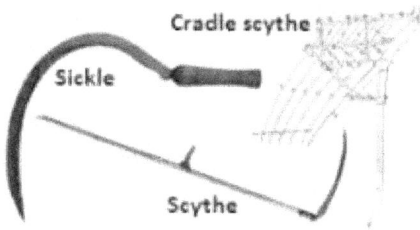

Figure 7-85: Reaper tools

Early reaping was accomplished by hand aided by such tools as mentioned above. Then the stalks were bundled into sheaves. These bunches of sheaves were leaned against each other so the sheaves could stand up. The standing bundles were called stooks.

Figure 7-86: Stooked wheat at Sommerfeld Colony, 1930

The stooks were left to dry in the field. Later, the sheaves were hauled to the barn ready to be threshed so that the farmer could sell it.

Reaping the harvest in early days

In Egypt a sickle with flint blade was used to cut crops. A sickle is a curved, hand-held agricultural tool used for harvesting grain crops. Mostly, they were very simple tools. Yet the Europeans continued to use the sickle until limited labour forced them to use the more efficient scythe.

In Europe the scythe had been introduced to crop harvest by the Romans. An additional feature that looked like wooden fingers that kept the grain flat until the end of the cutting swing was added perhaps as early as 1803. This tool is called grain cradle.

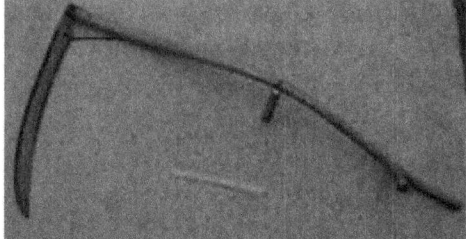 

Figure 7-87: A man using the scythe

Grain cradle scythe, (c.1830-1860)

The grain cradle is a scythe with four wooden fingers which gather grains while it is being cut by the iron blade. It was a very popular implement on the farm. Although developed as a result of a labor shortage, it did not save many hours of work as hands were still needed to collect and bind the grain. The grain cradle did have the advantage of being able to pick up and cut grain that had been beaten down by storms. It was made virtually obsolete by the mechanical reaping machine.

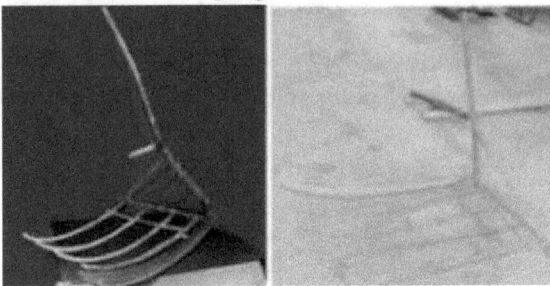

Figure 7-88: Grain cradles, (Left: c.1830-1860, Right: Today)

An original painting of a reaper with a cradle scythe produced by Ernst Henseler (1852–1940) is shown in the picture. Note the stooked harvest at the background.

Figure 7-89: Cradle scythe produced by Ernst Henseler

Another straw cutting device out of the ancient ingenuity and creativity was the chaff box cutter

Chaff (box) cutter

The chaff box or cutter was a simple but ingenious device for cutting straw chaff, hay, and oats into small pieces- before being mixed together with other forage and fed to horses and cattle.. The chaff box was the forerunner of all straw cutting machines and undoubtedly one of the most creative inventions in pre-mechanical agriculture. Although comparatively short-lived in Britain, the basic design had lasted for over four hundred years in Europe.

The chaff box was made largely of wood (usually ash) with only a small amount of ironwork, commonly found in farms, town or country stables by the end of the eighteenth century. In the early days some enterprising labourers and part-time thatchers learnt to use small portable chaff boxes which they carried on their backs visiting town and country stables offering a chaff cutting service for an agreed rate.

Figure 7-90: Three-legged chaff or monkey box

The barn version typically comprised of an open-ended, three-sided wooden trough between 3 foot 6 inches and 4 feet (1067-1220 mm) in length, 9 to 12 inches (229-305 mm) wide and 9 to 12 inches deep, mounted on either three or four legs approximately 22 inches (560 mm) high - giving an overall height of about 34 inches (863 mm). The rear leg or legs were sometimes taller than the front to tilt the trough forward to facilitate cutting. Both front legs of an early three-legged form were distinctly bow-shaped. Most of the chaff box's features had been established around the turn of the eighteenth century.

Figure 7-91: Small portable chaff box, (Blundeville, 1565)

Earliest known depiction of a chaff-cutter: woodcut from Augsburg, 1524 is shown in Figure 7-91 (Blundeville, 1565). Perhaps noting the early development of these rotary machines, attempts were made to improve the box still further. Despite, these developments, which peaked around the 1830's, the chaff boxes became increasingly displaced by the faster and then more efficient rotary cutters (which had first appeared in small numbers in the 1770's) with their automated feed systems.

Animal drawn reapers

Labour shortage, both in Europe and especially in the Western United States, spurred the farmers to find new and more efficient ways to harvest their crops (Grigg, 1974). Horse drawn mechanical reapers later replaced sickles for harvesting grains. The first attempts to build a machine to cut grain were made from the Gauls (England and Scotland) in Europe, in the 18th century.

Figure 7-92: Gaullic reaper header, c 1700

The first recorded English patent for a mechanical reaper was issued to Joseph Boyce of London in 1799 while another version was developed by Englishman Thomas James Plucknett, Deptford in 1805. By 1811, several people had taken a shot at developing reaping machine. Mr. Smith produced a horse-drawn machine that cut grain with a moving disc turning level with the ground. By 1814 another reaping machine was patented by Englishman James Dobbs. As early as 1820, horse-drawn reaping-sheaving machine were invented by a Englishman simply known as Brown.

In 1822, a school teacher, Henry Ogle, invented a mechanical reaper, but the opposition of the labourers in the vicinity, who feared loss of employment, prevented Ogle from making any further innovations. Four years later, the first successful reaper was constructed by Rev. Patrick Bell (1799-1869) in Angus, Great Britain in 1826. Patrick Bell, a Presbyterian minister, who had been moved by the hard work of the harvesters on his father's farm in Argyllshire, made an attempt to lighten their labour.

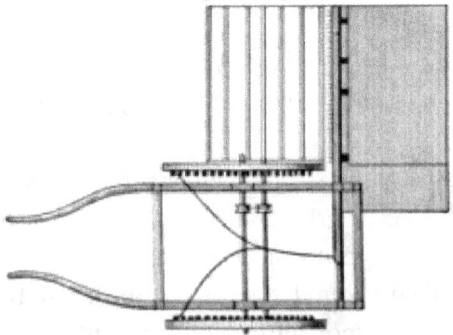

Figure 7-93: Ogle reaper

His reaper (Figure 7-93) was pushed by horses; a reel brought the grain against blades which opened and closed like scissors, and a traveling canvas apron deposited the grain at one side. In this design the reaper was pushed by horses with the shears cutting the wheat in front (Blandford, 1976). The Bell reaper (Figure 7-94 as illustrated in by Blandford) could cut ten acres a day and needed sharpening after fifty acres (Neilsoon, 1936).

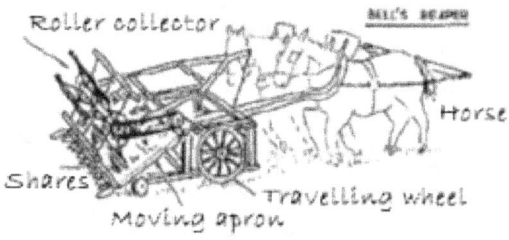

Figure 7-94: Bell's reaper

The inventor received a prize from the Highland and Agricultural Society of Edinburgh, and pictures and full descriptions of his invention were published. Several models of this reaper were built in Great Britain, and it was reported that four of his machines were exported to the United States; however, Bell's machine was never generally adopted.

Figure 7-95: Bell's reaper, 1826

Reaping machine based on the use of rotary blades was also developed by Patrick Bell, in 1826 but was not patented, while he invented and used reciprocating knife on one of his first successful reaping machines. The reaper used many reciprocating serrated triangular shaped knife blades spaced about two inches (5 cm) apart riveted onto a long metal rod 3 to 20 feet, (1–7m) long, the sickle bar cuts the grain. The sickle bar's, guard plates mounted on each side of each knife blade holds the grain while the reciprocating sickle bar's knives sheared the grain off. The reaper was pulled by a team of two horses walking at the side of the grain allowing the reaper to be pulled instead of being pushed by the draft animals.

The reaper driver and a helper that removed the grain into rows for later binding usually rode on the reaper. Improvements were introduced on the machine as the Beverly Reaper in 1826. 1828 witnessed the development of a cutting device based on scissors principles that formed part of reaping implement pushed in front of horses, built and demonstrated, but not patented, by Patrick Bell (1799-1869), of Scotland.

Figure 7-96: Hussey reaper machine

Soon afterward three men patented reapers in the United States: William Manning, Plainfield, New Jersey, 1831; Obed Hussey, Cincinnati, Ohio, 1833; and Cyrus Hall McCormick, Staunton, Virginia, 1834. Just how much they owed to Patrick Bell cannot be known, but it is probable that all had heard of his design if they had not seen his drawings or the machine itself. Bell claimed earlier invention of reapers but could not compete with the reapers of Cyrus H. McCormick (1809-1884), and Obed Hussey (1792-1860).

The first of these inventors, Manning of New Jersey, never made a machine other than his model. More persistent was Obed Hussey of Cincinnati, who soon moved to Baltimore. Hussey was an excellent mechanic. He patented several improvements to his machine and received high praise for the efficiency of the work. But he was soon outstripped in the race because he was weak in the essential qualities which made Cyrus McCormick the greatest figure in the world of agricultural machinery.

Figure 7-97: McCormick's reaper (From The Prairie Farmer, January 1941)

The invention of these two successful reaping machine inventors - independently by Obed Hussey in Ohio, who obtained the first patent, in 1834, and by Cyrus Hall McCormick in Virginia - brought about an end to tedious handiwork and encouraged the invention and manufacture of other labour-saving farm implements and machinery.

An improved grain cutting blade (a sickle bar with replaceable serrated triangular shaped teeth) acquired in 1850 made the reaper much more efficient. The long involved harvesting process was still there only the grain cutting was easier and faster. Bell reaper was designed by Scot Patrick A. Bell (1799-1869) in 1858, and was later built.

Figure 7-98: McCormarck-Deering reaper

Self-binding reaper was patented by Edward Sabine Renwick (1823-1912), with Peter H. Watson in 1851. Combined reaper and mower was also patented by John H. Manny (1825-1856), first developed by Manny & Co. in 1856 in Rockford, IL, and later by Walter A. Wood (1815-1892), who made improvements and commercialized the device.

Machine drawn (sail) reapers

The sail reaper was invented in 1862. It replaced a person pushing off the stalks by a sail-like rake. The rakes rotate and clean the stalk into piles off the side of the machine. Then men would again manually bind up the stalk into bundles (Blandford, 1976).

Figure 7-99: Sail reaper (McCormarck sail reaper on the left)

One of the largest used early sail reapers was made by the McCormack Company, widely used and accepted in United States and England. The rest of Europe was much slower to adopt the new technology. A later improvement called reaper-binder that cut the grain and tied it into bundles ready for collection and harvesting showed up about 1870.

Reaper-binders development

The reaper was improved to both reap and bind the stalks in later versions by McCormick and other manufactures. McCormack started mass-production of the reaping machines which bound the sheaves in 1877 (Lee, 1960). Cyrus McCormick had many competitors, and some of them were in the field with improved devices ahead of him, but he always held his own, either by buying up the patent for a real improvement, or else by requiring his staff to invent something to do the same work.

Reapers developed into and were replaced by the reaper-binder (which cuts grain and binds it in sheaves), invented in 1872 by Charles Withington, which was in turn replaced by the swather and then the combine harvester.

Harvesters

The earliest known records of the first harvester, was the often mentioned Gallic stripping header described by Pliny in the first century and by Palladius in the fourth. It appears to have been uninterruptedly used for several centuries, and unquestionably it had not been in use that long without having been more or less improved, for the people who could invent and construct such a machine would surely improve upon it, and they would also invent others of like character. There is no doubt that various reaping and harvesting machines were used by the ancients of which we have no record, principally because agricultural pursuits were not honored and historians gave their attention chiefly to matters of government and war.

The advent of the Marsh harvester in 1789 by William W. Marsh (1836-1918) and his brother Charles W. Marsh (1834-1918), in Shabbona Grove, Illinois marked the beginning of the more frequent use of the term "harvester" which had been applied almost exclusively to any particular kind of machine which carry binders, whether men to bind by hand, or automatic binders, substituted for the men, to bind mechanically. But this use of the term is arbitrary and narrow; it should take in strippers, headers and combined harvesters and threshers; and it will at least be better for our purpose to give it a wider meaning.

Figure 7-100: Marsh harvester, as built in 1879

The Marsh harvester has a binder attached to a reaper that gathered grain in bundles. A still later improvement used a type of conveyor belt behind the reel to load the gain and stalks directly into a wagon travelling next to the reaper. The wagons hauled the cut grain to the thresher where the gain was separated from its stalk and chaff. The horse pulled reaper had to stop until a new wagon could be driven up to collect the cut grain.

Figure 7-101: Woodcut of an early Marsh brother's harvester

The success of the Marsh's design was owed to the fact that during demonstration anyone could use the implement. The system involved movable canvases coupled with a platform and table that allowed one or two men to ride a reaper while tying the cut grain into bundles. Before that innovation, cut grain was raked off the reaper in gavels, (or straight piles) and then men who followed the machine, gathered each gavel and tied it into bundles.

Figure 7-102: Grain binder cum thresher (Northern Great Plains, 1880-1920)

Numerous new devices to improve the harvester were developed, history of which spanned over a century but the most important was an automatic attachment to bind the sheaves with wire which was patented in 1872 by Cyrus McCormick as his own. The harvester seemed complete. One man drove the team (Figure 7-94), and the machine cut the grain, bound it in sheaves, and deposited them upon the ground. In 1890 only one tenth of France or Germany had adopted the use of the reaper in their fields (Grigg, 1974).

Figure 7-103: McCormick corn binder

The main complaint about the first harvesters was about the wire ties. When the wheat was threshed, bits of wire got into the straw, and were swallowed by the cattle; or else the bits of metal got among the wheat itself and gave out sparks in grinding, setting some mills on fire.

Two inventors, almost simultaneously, produced the remedy. Marquis Gorham, working for Cyrus McCormick, and John Appleby, whose invention was purchased by William Deering, one of McCormick's chief competitors, invented binders which used twine. By 1880, the self-binding harvester was complete.

Twine binder

The invention and development of self-binders was progressive with the era began with the adaptation of self-binders to the Marsh harvester and runs through the various stages of development in wire and twine binders, and down to the time of the third, or current era, wherein all useless and impracticable devices have been eliminated, all material for bands has given place to twine, and all machines upon the market, with one or two exceptions, are substantially alike in form and general principles.

John E. Heath, of Warren, Ohio, was the first of record to attempt to bind grain by machinery, and his was a twine or cord binder. Patent of device for twine binding of sheaves (binder) manually for reaper was awarded John E. Heath, Warren, OH. His patent was dated July 22, 1850. But little is known of Heath's binder except from his patent. It is said that he built several machines and that they operated fairly well for a first effort.

The next patent for a binder was granted to Watson, Renwick & Watson, May 13, 1851. This patent is a curiosity and a study. The specification is exceedingly long, with many drawings, and it is reinforced by two or three pages of modifications, the inventors evidently intending to cover every form of binding device that they could think of.

Geo. Wood N. Yost, of Mississippi, Jan. 1, 1856, obtained a patent on a machine for binding with a cord band, cut to right length, with a knot tied in one end. This knotted end was placed in a notch and the other end went somewhere. A gathering and compressing apparatus swept along the platform, forming a bundle, around which the band was brought, when its ends were tied by an attendant.

Wood's invention was not automatic and they added little or nothing to advancement. The next machine showed decided progress. This was the invention of C. A. McPhitridge, of St. Louis, Mo., for wire-tying binder, hand operated by whom it was patented Nov. 18th 1856.

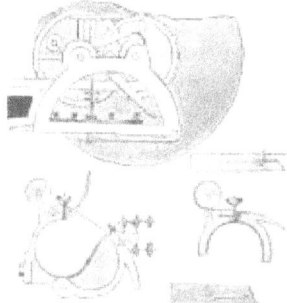

Figure 7-104: First wire binder, M'Philtridge patent, 1856

A self-tying device to bind sheaves of straw with twine was developed by John F. Appleby (1840-1917). A twine tying knotter reaper invented in 1858 by John F. Appleby, was not developed further because of the cost of twine.

Figure 7-105: The Appleby experimental binder, as made at Beloit

March 2, 1858, W. L. Childs, of New York, patented an ingenious twine-binder. The cord was taken from a spool located in the grain-wheel divider. It was passed under the platform and around in front of the receptacle into the nippers, above, in the arm; a self-rake swept the grain against the cord, which was forced back, receiving and encompassing the gavel; then the arm came down, the twine was cut off, and the ends were twisted and tucked under automatically.

On August 17th 1858, C. W. and W. W. Marsh patented their hand-binding harvester, which, though not belonging specifically to this class became finally the foundation upon which practical binders up to this time have been built.

To J. Mitchell, on Sept. 7th 1858, was granted a patent for an automatic straw-binder. This invention consists in the use of clamps or band-carriers, a band-twisting device, tucking rod, and discharging device applied to the reaper, arranged relatively with each other and operated, whereby the grain is bound into sheaves and discharged upon the ground, the entire system working automatically as the machine moves along. It was not practical. The next patent, Nov. 16, was to Wm. Gray, of Ohio, covering ingenious contrivances for binding automatically with straw.

By 1863, wire binder for reaper were built in large numbers and adapted to Marsh harvester in 1872. In the long list of meritorious inventors the name of James F. Gordon, of Rochester, N. Y., and of his brother, John H. Gordon, should stand out prominently on account of their valuable work and their persistent efforts. James F. began inventing in this line as early as 1862 and had a full-sized machine in 1864. He continued his experiments under adverse conditions

There were several binder patents granted during the year 1865, but none of them represented successful inventors or machines, except two issued Dec. 19, to S. D. Locke, of Janesville, Wis., one covering a compressing device, the other his rotating hook twister. Locke says he began in 1861 to build a binder which, after nearly completing, he abandoned, to commence on one of another style or plan.

Figure 7-106: S. D. Locke's wire binder, 1873

Mr. Locke took out patents too numerous to mention. He claims to be the first man to build an automatic binder as a distinct and separate machine for attachment to a harvester, and Walter A. Wood & Co. were the first to build and put regularly upon the market successful automatic binding machines. While it is probably true that Mr. Locke was the first inventor of binders on record who made a final success, and that success began with the adaptation of his binder to the Marsh harvester in 1872.

Twisted wire used for fencing was patented in 1867 (but did not include method of manufacture) by Lucien Smith, of USA. Patent for another type of barbed wire for fencing was issued the same year to William D. Hunt, Scott County, NY, USA., a wire-tying reaper binder was patented by American Walter Wood (1815-1892) in 1871.

Figure 7-107: Wood's binder

Between 1871 and 1873 Wire-tie balers were developed by Sylvanus D. Locke (1833-1896), followed by self-tie binders in 1875. A wire-tying binder patented by John F. Appleby (1840-1917) was changed to twine tying in 1874, at which time he formed the Appleby Reaper Works, followed by its commercial use on the McCormick binder.

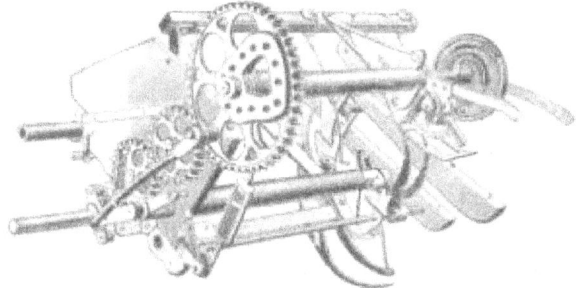

Figure 7-108: The Appleby standard binder, as first put on the market

Automatic wire-tying devices were also developed and in production in 1873, made to fit the Marsh Harvester by John H. Gordon.

Figure 7-109: John H. Gordon's parker binder, 1873

Subsequently, machine capable of producing barbed wire in large quantities patented by Joseph F. Glidden (1813-1906), De Kalb, IL, appeared same year, followed by commercially successful production. Glidden worked with Isaac L. Ellwood (1833-1910), who also patented barbed wire for fencing. During this era several patents advanced for barbed wire fences, followed by numerous legal challenges.

Twine knotter to be attached to binder to tie cut grain into bundles was patented by John F. Appleby in 1878 (see Figures 7-105 and 108), and adapted to Marsh harvesters (Figure 7-100). Appleby tying devices were used on binders made by William Deering (1826-1913) from 1878.

Figure 7-110: Deering corn harvester

The Appleby-type twine-knotter binder invented by John F. Appleby, supported by William Deering in 1877, was launched by McCormick in 1881, and licensed to other companies. In 1883, McCormick switched from wire binders to twine binders, advanced by Cyrus H. McCormick, Jr (1859-1936), starting intense competition with William Deering & Co. By 1892, self-binding corn binders were patented.

Beyond this period, several inventions and development in binders' technology evolved with John Deere Co. and Messey-Harris Co playing major roles by leading the revolution into the era of industrial agricultural production.

Figure 7-111: John Deere binder

Figure 7-112: Massey-Harris binder

Animal treadmills

J.A. and H.A. Pitts patented a horse treadmill in 1834. "Pitt's endless chain and cogband" featured iron chain links and many hardwood rollers to support the entire tread and prevent it from sagging. They manufactured this treadmill and also a sweep horse power to power their groundhog threshers.

As treadmill development continued after the 1830's, one important improvement was the "level tread" design, where the treads remained horizontal, rather than sloped upward. This provided a surer footing for the horse, and less leg strain. The Heebners of Montgomery County, PA, first patented a level tread for their toothed chain treadmill in 1871. The triangular links allowed an inclined lower surface and a horizontal upper surface, resembling a mini-escalator. In 1883, the Heebners received another patent for an improved level tread, now using iron cross rods to secure the links on opposite sides of the track, rather than relying on the tenoned wooden treads to do this.

Threshing

Threshing is the process of separating kernels of grain from the hull (the hard outer covering) and straw so that the grain can be turned into threshing flour. The threshing floor, on which oxen or horses trampled out the grain, was still common in George Washington's time. The grain was spread out on the floor of the barn and hit with a tool called a flail.

Figure 7-113: Flail

Flail: The flail was a long stick attached to a wooden handle. The flail threshed the stook while seeds, chaff (bits of seed head) and straw remained after flailing. After most of the straw was raked away, the farmer gathered what was left with grain shovels.

Figure 7-114: Sweeping the threshing floor in order to pile up the seed

Shovel: Grain shovel (scoop shovel) is carved from a single piece of wood with common dimension (49"L x 13 3/4"W (scoop 14 1/2"L, handle 34 1/2"L). Such wooden shovels were widely used for moving grain as early the 19th century farmers believed metal would bruise the seed.

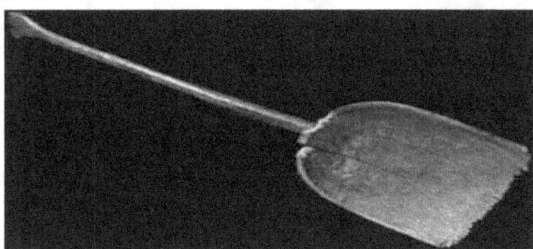

Figure 7-115: *Grain shovel (c.1800-1850)*

Wheelbarrow: These are small carrying aids that have wheels and are used for loading and unloading agricultural material, whether sand, soil, fertilizers. The wheelbarrow was first invented in the Ancient China by Chuko Liang (181-234 AD). Liang was a general who used the wheel barrows to transport supplies to injured soldiers. The Chinese wheelbarrows had two wheels and required two men to propel and steer It was probably independently invented in Europe in the 12th century.

Figure 7-116: Ancient Chinese wheel barrows

Threshing process

When grain was being cut by hand, the method for separating the kernels from the straw was slow and labour intensive. Grain was hauled to a barn where it was spread on a threshing floor and either beaten with hand flails or trampled by animals. That knocked the kernels free of the straw, which was then raked away.

The remaining mixture was winnowed by tossing it into the air where the wind was relied upon to blow the chaff and lighter debris away from the heavier grain, which fell back onto the threshing floor. These processes are not only labour intensive, but slow and hazardous. The invention of threshing machines both pulled by animals and later by engines transform agricultural crop processing tremendously

Threshing machine

For thousands of years, grains were separated by hand with flails, a very laborious and time consuming and taking about one-quarter of labour throughout the 18th century and in early 1777, scythes, sickles, hoes, spades, shovel, and millstones had already found ready market in USA.

Figure 7-117: Roman harvesting machine

The origin of the threshing machine was attributed to Scot Michael Menzies (?? -1766); threshing machine (in 1732) and flail threshing machine (in 1734), however, some historical records attributed the first invention of threshing machine, which in earlier years were powered by horses, to a brilliant Scottish mechanical engineer Andrew Meikle (1719-1811) in probably 1784.

Figure 7-118: Horse drawn thresher

The early machines were hand fed and powered by horses and later by portable or traction steam engines. Most of these massive machines were stationary and did many different things. First it removed grain, and then separated the grain from the cob or husk, after that it cleaned the grain and then gathered and stacked it. In Great Britain several threshing machines were devised in the eighteenth century, but none was particularly successful. They were stationary, and it was necessary to bring the sheaves to them.

One patent issued by the United States to Samuel Mulliken of Philadelphia, was for a threshing machine. In 1786, rotary drum type threshing machine was developed by Andrew Meikle of East Lothian. It used a fan to separate chaff from grain and became a standard for this operation. Threshing machine appeared in America through an American inventor, Isaiah L. Jennings in 1810.

Figure 7-119: Batteuse Damey animal powered threshing machine, 1881

Steam powered threshing machine emerged in 1812 invented by Englishman Richard Trevithick (1771-1833). The Pitts brothers — Hiram A. Pitts (1799-1859) and John A. Pitts (1799-1859), of Winthrop, Maine, were the first successful American inventors whose inventions went into practical and general use until recently.

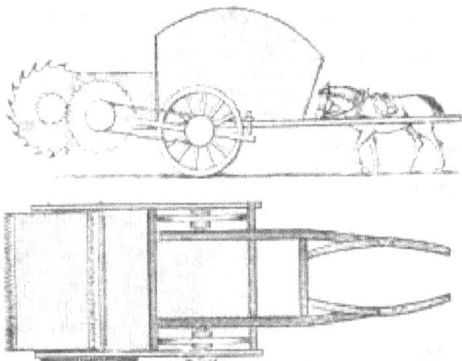

Figure 7-120: Pitts' machine, 1837

Pitts built a steam powered combined thresher-winnower in 1837 and also introduced steam-powered mechanical thresher and fanning mill same year.

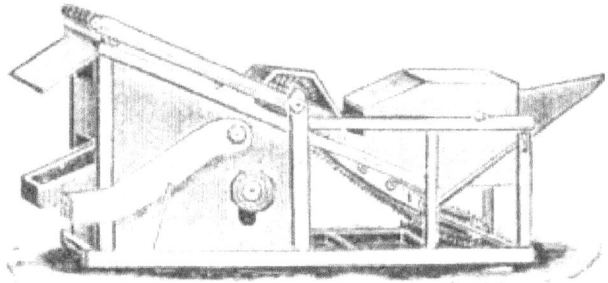

Figure 7-121: Pitts' thresher, 1838

The year 1850s saw the development of transportable threshing machine that could be taken to the fields developed by Englishman Tasker and also the introduction of shaker or vibration principle employed to separate grain from the straw for some harvesting and threshing machines.

Doubtless several kinds of fairly practical powers were constructed and used before any showed sufficient superiority to create a type, and probably they were nearly or quite all of the low speed, that is, requiring jacks to give sufficient motion; but as with the "separator" so with the power, it seems that the Pitts brothers were the leaders. At any rate the Pitts power was the first to gain general use and to maintain its position, as improved, of course, in the market down to the present day.

Figure 7-122: The sweepstakes thresher, Pitts' system

Powers designated by their motions are of two classes — the low and the high speed. The first requires the jack to increase motion, and the second furnishes sufficient speed direct. These powers for threshing machines are now obsolete, but for other purposes are very useful, and are generally manufactured. Of the early high-speed powers the planet, the woodbury, the triple-gear and the climax were the leaders. The first lever-power used by the Pitts brothers had the large master-wheel and single pinion with jack; but they were experimenting upon the climax, so-called, about 1845, when Mr. Carey, who was working in the same room with H. A. Pitts, suggested using two pinions instead of one for the two bevel-wheels in the center to prevent heat and wear.

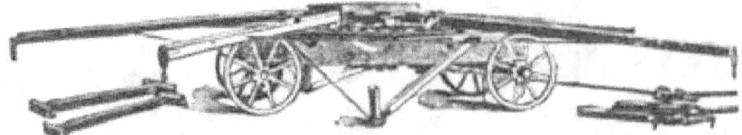
Figure 7-123: Pitts-Carey power, mounted system, 1845

The result was the internal gear and the turning of the bevel-wheels down, and thus the foundation of the Pitts-Carey power was laid. Mr. Carey made no claims to the invention at that time, but the Pitts brothers, in honor of his suggestion, had his name signed to the application for patent, taking assignment in full; and afterwards they paid him $500, which he thankfully received.

Stationary axial-flow threshers were patented by Gregor & Gregor in 1886 in Germany. Red River Special, a popular and successful thresher, was introduced by Nichols & Shepard Co., Battle Creek, MI (John Nichols) in 1905. Centrifugal threshing machine was invented by Thomas Forster, 1870-1946 in 1917. Equally there exists record of G. F. Nye having a patent for centrifugal thresher with no date listed.

Figure 7-124: Case steam threshing at Perdue, SK, 1920-1925

The threshing machine was never widely purchased by small farmers. They were very large and often too expensive for the average farmer. Often they were used in custom operation going from farm to farm. Later these threshing machines were combined with reapers and became known as combines.

After various improvements had been devised, significant improvements to a machine that automatically threshes and separates grains from chaff, freeing farmers from slow and laborious process were invented. E. W. Rowland-Hill developed the rotary threshing concept in 1943.

Combination harvester and thresher

Another development grain processing was the combination of harvester and thresher in one unit as seen in use on larger farms of the West. This machine does not cut the wheat close to the ground, but the cutter-bar, over twenty-five feet in length, takes off the heads. The wheat is separated from the chaff (threshed) and automatically weighed into sacks, which are dumped as fast as two expert sewers can work.

Winnowing practices

Early winnowing practices involved the grain seeds and chaff mixture placed in a winnowing tray (or basket) and shaken and tossed on a windy day. The wind would blow the light straw and chaff away and the seed would fall back into the tray. This was called 'winnowing' and was done over and over. This process could take up to two months. Winnowed grain was stored for animal feed or taken in sacks to the mill to be ground into flour. Figure 7-125 depicts a Jean-Francois Millet picture showing a farmer winnowing grains.

Figure 7-125: A farmer winnowing grain

The origin of winnowing implements and their modest development is uncertain. The first winnowing machine recorded in history was developed by James Sharp (according to dated engraving) in 1770.

Figure 7-126: Winnowing machine courtesy of Roy Bridgen

Certainly they were used quite extensively in central and Southern parts of England and Wales during the 17th, 18th and early 19th centuries before being gradually replaced by the internal fan bladed winnowing machine. As an agricultural implement it was not confined to Britain since other European examples existed at the same time particularly in Spain and Italy.

Early attempts to create an artificial wind to winnow corn instead of waiting for a natural wind to blow through opened barn doors resulted in a device variously called a *winnowing fan, winding* or *sail fan*. In 1669 "Dictionarium Resticum" defined a 'fann' as "an instrument that by its motion artificially caused wind. Due to their whirling action, some were provisionally called *gigs* or *ginning* machines, names that were also given to machinery designed to teasel cloth.

Figure 7-127: Doukhobor women winnowing grain, 1899

Other names included fan or van which additionally referred to winnowing baskets. The simplest and probably the oldest form comprised of three or four flaps of cloth or leather nailed directly to a hand-cranked spindle mounted horizontally on an 'A' shaped trestle. While one man whirled the fan around to deliver a strong and dependable stream of air, another using a wooden shovel would toss the gain into the draught to blow the chaff away.

Larger improved versions appeared, one having three or four widths of Hessian or canvas sail cloth attached to a corresponding number of supporting frames or spars which were turned by a sturdy, end-mounted, wooden-handled wheel.

Screens

Screens were used in Tusser's time during the 16th century. In 1825, Loudon reported that the screen was chiefly used in granaries to free corn from weevils, while in 1933 Hennell noted that millers in Kent knew it as a *Scry*. Those in Sussex and Surrey however called it a *Scroy* but elsewhere its earlier similarity to the musical instrument caused it to be known as the *Corn Harp*.

Winnowed corn or grains may still contained impurities such as weed seed, insects and dust which needed to be removed. Smaller amounts were re-cleaned by hand held sieves but greater quantities usually necessitated the use of large freestanding corn screens to sift out these impurities before it was sent to the market or the miller.

Figure 7-128: Wooden hand sieve

The original form of corn screen comprised of a simple oblong frame some five to six feet (1524–1829 mm) in length by two and half to three feet (762–914 mm) in width. The frame held a series of closely spaced longitudinal wires and stood at an angle of approximately forty-five degrees supported by two back legs. The grain was tossed at the upper screen with a wooden corn shovel (called in parts a scuppit). The grain rolled down the front into a winnowing basket or receptacle placed below while the smut fell through the wires to the back.

A number of similar designs were patented. One by J Francis in 1864 contained three roller brushes positioned down the frame at intervals which were rotated by hand to polish beans, oil and scour seed and to remove the 'web from peas' and 'mites from turnip seed'. Another in 1875 had a number of adjustable inclined shelves or cross slates like a step ladder over which the grain jumped the open spaces while the debris fell through. By this time however Boby's of Ipswich had already introduced their patented 'self-cleaning and self acting corn screens' along with other manufacturers in that period including Corbetts Hornsby & Sons, Denny & Co and W Rainforth & Sons. Most were hand operated and gradually gave way to powered machinery.

Hay harvest and forage machine development

The hay harvest also benefited from mechanization. Just as ancient man came up with the idea of inventing the wheel, it was probably only a matter of time until someone devised the idea of squeezing loose hay into a package that could by tied, handled and transported. But until the mid-1800s, hay that was harvested for livestock was simply piled into stacks or moved into the barn for use during the dry season.

Long and before the invention of hay and forage machine in the middle of the 19th century, hay was cut by hand with sickles and scythes, while pitchforks were used to pick up, move and stack hay and straw. Moving the crop involved pitching it onto a wagon and pitching it back off at the destination using some hand-held devices such as pitchfork.

Pitchfork: Pitchfork was made from a single piece of hickory at the Rush farm in Fairmount, Ind., sometime between 1840 and 1860. One end of the board is sawn into quarters and each quarter is shaped down to a pointed tine. The four tines are separated, secured and strengthened by a wooden dowel, iron brads and iron rivets. The long handle was a great help in building stacks and the curved tines made it possible to pick up and hold the material.

Figure 7-129: Pitchfork, c.1840-1860

In the 1860s early cutting devices were developed that resembled those on reapers and binders; from these came the modern array of fully mechanized hay-making implements, mowers, crushers, windrowers, field choppers and balers.

Hay-making implements

Hay tedder or kicker: In the early 1800s, a hay-making implement called tedder or kicker for hay was developed by Robert Salmon (1751/52-1821), Woburn, England to turn grass and forage. Salmon introduced a device for cutting forages consisting of an oscillating clipper with a row of smooth shear blades in 1807.

Hay rake: The first horse-drawn revolving hay rake patent was obtained by Samuel Pennock in 1823; however, hay rake development had already became popular in 1820.

Figure 7-130: Horse drawn rake

The first hay rake manufacturing business was established by William Stedman, which greatly expanded production of hay rakes and other implements by his son Marshall W. Stedman (1859-1935) in 1827.

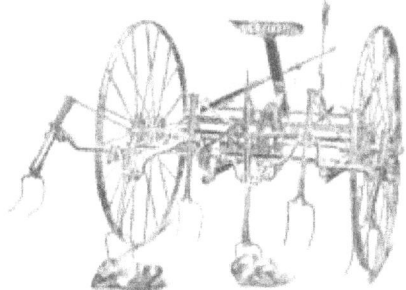

Figure 7-131: Osborne all steel hay tedder

Hay tedders that turned cut grass developed by Englishman Robert Salmon in 1820; were replaced by side-delivery rakes at the beginning of 1830s. Mr. Wedlake, of Essex, England improved the tedder to lift and toss the hay in 1830.

Figure 7-132: Walter A. Wood side delivery rake

In 1838, Horse-drawn hay rake was invented by General Lewis Swift (1784-1846) and manufactured by his son Lewis Swift Jr (1820-1913). Self-raker machine called Atkins Automaton Raker was introduced to the public in 1841, and patented in 1853, by Jearum Atkins (1840-1880), with 5000 units sold in 1856.

Figure 7-133: The beck side delivery rake

Figure 7-134: The New York self rake

By 1850, hand-dump hay rake made of iron or steel teeth was introduced to the public. The innovative Wood's self-rake reaper was introduced by Massey-Harris-one of the two MF forerunners in 1861. By 1889, Haseley tedder or kicker for hay manufactured by a number of British firms; were soon to be replaced by the swath turner and side-delivery rake. Field hay chopper patent was issued to William J. Conroy in 1891 and reel-type side delivery rake was introduced in 1914.

Figure 7-135: Modern hay rake

Hay and forage bailers (presses)

The hay harvest also benefited from mechanization. Just as ancient man came up with the idea of inventing the wheel, so it was only matter of time until someone devised the idea of squeezing harvested loose hay into a package that could be tied, handled and transported. Until the middle of the 19th century, hay was cut by hand with sickles and scythes. Moving the crop onto a wagon and pitching it back off at the destination all that changed in the mid-1800s, with the invention of the first mechanical hay press.

Stationary bailers: Most of the earliest hay presses were stationary units built into a barn and extending two to three stories into the hayloft. Generally, there are different versions of the stationary bailer units developed at that time. In one version, a team of horses was used to raise a press weight, which was then dropped to compress the hay. Other versions used a horse- or mule-powered sweep at the bottom of the press to turn a jackscrew or a geared press.

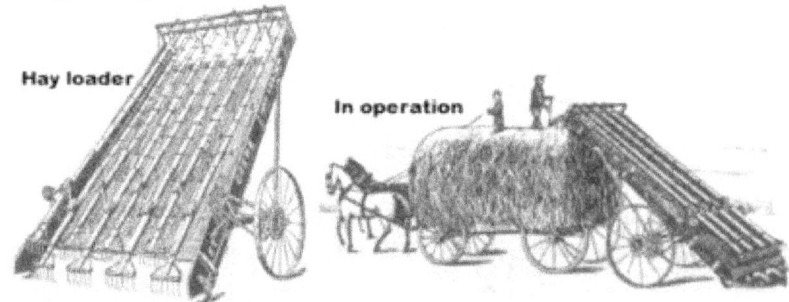

Figure 7-136: John Deere loader in operation

Also, workers are seen pitched loose hay into the baling compartment, where a hinged door opened to the side of the press. Once the compartment was filled with hay, the door was closed by counterweights. The attendant then pulled the trip lever, which allowed the weight to drop into the baling compartment and compress the hay. It usually took six or seven cycles to form a 300-pound bale.

The press also included a jackscrew, which pushed the baling compartment floor downward when the mule was led counter-clockwise to lift a 1,000-pound wooden weight to the third-floor level via a pulley. However, to finish the bale, the mule was led clockwise six times, which rotated the jackscrew to again bring the bottom of the bale level with the second floor. At that point, the door was opened, the bale tied and removed.

Unlike later hay presses, these permanent models often made bales weighing as much as 300 pounds, secured by as many as five strands of wire or twine. One such press was built by P.K. Dederick's Sons, Albany, N.Y., in 1843. Another, invented in 1843 by Samuel Hewitt of Switzerland County, Ohio in the United States of America, was seen on display at a landscape company in Lawrenceburg, Ind. Marketed as the Mormon beater hay press, powered by a mule attached to a sweep at the bottom of the press.

The stationary baler or hay press was invented in the 1850's by an American, Horace L. Emery (?? -1892), of Albany, NY precisely in 1853 and did not become popular until the 1870's. The baler (press) produced bales weighing approximately 250 lb. Baler enhanced by Peter K. Dederick (1838-1911), with continuous production was introduced in 1872, followed by many improvements including windrow hay loaders introduced by Keystone Manufacturing Co., Rock Falls, IL, same year. Steam-powered continuous-production balers were introduced in which bale was ejected automatically and manually tied with wire in 1884.

Portable (mobile) bailers: It wasn't long before hay presses became more mobile, going to the field and from farm to farm, much like the threshing machines of the day. Consequently, the first people to own the machines were custom operators and hay dealers who would buy a quantity of hay from a farmer then bale it before transporting it to market. A few models, however, were sized and built for private use.

Like the larger models on the market, those required the hay to be deposited into a square chute and compressed. The difference was that these models used manpower (via a lever or crank) rather than horsepower. One such baler, on display at the Pioneer Village in Minden, Neb., was built by William Henry Penniston of Fox, Mo., in 1863. The wooden hay press was entirely hand-operated, requiring workers to fill the upright chamber with hay, pack it with a lever-operated jack-type ratchet, tie the bale with wire or cord once the chamber was full and then open the front door to eject the bale. With a coordinated effort, a crew could supposedly build up to 72 bales in one day.

Hay conditioners and pick-up balers: Hay conditioners were invented in 1912 by a German named Hermann Bartsch while Ann Arbor baler was considered the world's first pick-up baler, introduced and manufactured by the Ann Arbor Manufacturing Co.

Figure 7-137: A 1913 Rumely hay press

Figure 7- 137 shows a 1913 Rumely baler as part of a Rumely collection owned by Jesse Boller, Ashland, Neb. Note the block of wood in the chamber that is used to separate and tie each bale. In the 1930s mechanical hay balers were at work, but the process still required hand tying of the bales. Jesse is seen in Figure 7-141 demonstrating how wire is fed through passageways in wooden blocks to tie bales on his vintage 1913 Rumely baler.

Figure 7-138: Demonstrating wire fed on a 1913 Rumely hay press

Between 1932 and 1935, Hume-Love floating cutterbar and pickup reel patented in 1932 followed by improvements, particularly the pick-up reel by Horace D. Hume and James Edward Love who formed a partnership (Hume-Love Co., Garfield, WA) in 1930; designed particularly for harvesting lentils, pea beans, dry peas, and soybeans but used more widely Commercially successful pick-up forage harvester developed by Erwin W. Saiberlich and manufactured by Fox River Tractor Co., Appleton, WI. Earlier models had been designed by Floyd W. Duffee, University of Wisconsin, and William J. Conroy, Ayllmer, Quebec, Canada.

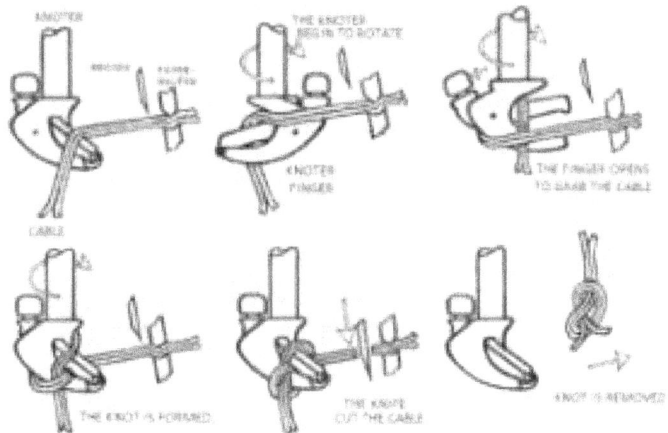

Figure 7-139: Appleby's knotter mechanism

In 1936, a man named Innes, of Davenport, Iowa, invented an automatic baler for hay. It tied bales with binder twine using Appleby-type knotters from a John Deere grain binder.

Figure 7-140: Behel's tying bill: forming the knot

In 1938 a Pennsylvania Dutchman named Edwin Nolt invented a machine that automated bale tying (Figure 7-141). Note that it still fed hay into the top of the chamber and used an overhead plunger much like those on early hay presses.

Figure 7-141: Ed Nolt's bailer on field test, 1938

According to the history of twine, Nolt's innovative patents pointed the way by 1939 to the mass production of the one-man automatic hay baler. His balers and their imitators revolutionized hay and straw harvest and created a twine demand beyond the wildest dreams of any twine manufacturer.. Ed Nolt built his own "pick up" baler or square baler, salvaging the twine knotters from the Innes baler. Both balers did not work that well, however, Ed Nolt built about 35 units of his field pick-up, self-tying hay baler in a shop near Kinzers, Pa., before he sold the design to New Holland Manufacturing Company which incorporated it into a pickup baler that it began marketing in 1941.

Figure 7-142: The New Holland Model 73 automatic pick-up baler

The first successful automatic pick-up, self-tying hay and straw baler manufactured by New Holland Machine Co., New Holland, PA in 1940. Development was based on work of a farmer in the area and engineers at New Holland Machine Co.

Round hay balers: In the 1940s, the "pick up" baler or square baler was soon replaced by the round baler first patented by Hugh Luebben and his sons Melchior and Ummo, Sutton, NE in 1903. By 1970, giant, round hay balers were invented and patented by Wesley Buchele and Virgil Haverdink, Iowa State University, Ames, IA and in 1972 major manufacturers of balers;the Vermeer Manufacturing Co., Pella IA., made large round forage bales weighing up to 2000 lb. A baler making small round bales for forages was built in the 1930s by Allis-Chalmers Co. and in 1966 Deere & Co. introduced a hay cuber.

Figure 7-143: Square and round bales

Combines (harvesters)

Combine harvester (combines cutting and threshing) patented by Samuel Lane of Maine appeared in 1828 but not operational until 1836. 1834 saw the first operational grain combine harvester developed by Hiram Moore (1801-1875) and John Hascall. The machine was patented in 1836, driven by a ground wheel and pulled by a team of 20 horses, built in Climax, MI. The combine was transported to California for demonstration and testing.

The first marketable combine was invented by Hiran Moore in 1838. In 1839, combine harvester similar to the Moore-Hascall unit was developed in Australia by H. V. McKay. Head-stripping cereal harvester, called the Australian stripper, invented by John W. Bull in 1843 was produced by John Ridley (1806-1887), in Adelaide, Australia same year.

Figure 7-144: Australian combined stripping harvester

Early combines were driven by as many as 16 or more horses. As its name indicated, it combined the two main tasks of grain harvesting: reaping, or cutting the stalks, and threshing, the process of separating the kernels of grain from the rest of the plant and then collecting the kernels. Early combines were pulled by large teams of horses and proved about as unwieldy as the first steam-powered tractors.

Figure 7-145: Early combines pulled by horses

Later they were pulled by steam engine and then combined into a single machine by George Stockton Berry, towed by the powerful new diesel tractors of the 1930s and taking their power off the tractors' engines. Berry took the straw and used it for fuel to heat the boiler. The header (or cutter) was over forty feet. This machine could cut and thresh over one hundred acres in a day Olney, 1984). The cost of the machine was also cheaper than the horse drawn reapers and stationary threshers. The cost of the reaper and thresher was about $3 a acre while the combine was between $1.50 and $1.75

Figure 7-146: Horse drawn combine in operation

Self-propelled combines

By 1886, operation of world's first self-propelled (using straw-burning steam boiler) combine was patented in 1887 in USA) with a PTO using steam from the traction engine to drive the harvesting mechanism, by George Stockton Berry, 1847-1917, was built and operated in Lindsay, Tulare County, CA. The side hill combine era was ushered in by Stockton Wheel Co., CA, by Holt Brothers, a revolution led by Benjamin Holt (1849-1920) in 1891. A corn silage harvester was patented by William J. Conroy in 1891 while an axial-flow grain harvesters was patented by Felix Schlayer in Germany in 1922,

The first self-propelled combine was developed in Australia in 1938, incorporating tractor and harvester in one, and improvements have been steady ever since. Commercially successful self-propelled combine harvester for small grains, the Massey-Harris M-H 20, was developed by an Australian engineer Thomas (Tom) Carroll, c. 1888-1968 in 1938. It was first used in Argentina, manufactured by Massey-Harris Co. using a 65 hp Chrysler engine that became a leader in the field. When completed the development of the Massey Harris 20, over eighty years ago, little did he knew from that day forward that Massey Ferguson self-propelled combines would become the world standard for combine design. Decades later, Massey Ferguson 9005 Series combines represent yet another milestone in "Bringing in the Harvest."

Deere & Co. first company claimed to have successfully market a corn-head attachment for combine, John Deere Models No. 45 and No. 10 in 1954. 1n 1955, Allis-Chalmers Co. phased out the All-Crop harvester models in favour of the Gleaner trademark upon purchase of the company, credited to Curtis C. Baldwin, 1888-1960, in Kansas and IHC produced its first combine with a corn head, applied to the Model 141H in 1956. In 1965 axial-flow machines for corn combines were introduced to the market by several manufacturers including International Rice Research Institute, Massey Harris, and a number of Southeast Asian companies, three French manufacturers such as ABM Rivierre-Casalis, Australian Walsh Maize Header in Queensland, International Harvester Co., and New Holland Co.

1975 saw the following models of combine harvesters with axial-flow threshing introduced to the market: by International Harvester Co.; models 1440, 1460, and 1480, and the Sperry New Holland TR70, New Holland, PA. Development started in 1962. Today, the most impressive of these grain-handling machines can cut swaths more than 30 feet wide, track their own movements precisely through global positioning system (GPS) satellites, and measure and analyze the harvest as they go. They are in no small measure responsible for a 600-fold increase in grain harvesting productivity.

They did it all: cutting, threshing, separating kernels from husks with blowers or vibrating sieves, filtering out straw, feeding the collected grain via conveyor belts to wagons or trucks driven alongside. This moving assembly line turned acre upon acre of waving amber fields into golden mountains of grain. It took several decades before the combine came into wide use.

Mechanical harvesters

The corn combine

The same basic combine design worked for all grain crops, but corn required a different approach. In 1900 corn was shelled by hand, the ears were thrown into a wagon, and the kernels were shelled by mechanical device powered by horses. The first mechanical corn picker was introduced in 1909, and by the 1920s one- and two-row pickers powered by tractor engines were becoming popular. Massey-Harris brought the first self-propelled picker to the market in 1946, but the big breakthrough came in 1954, when a corn head attachment for combines became available, making it possible to shell corn in the field. The increase in productivity was dramatic. In 1900 one person could shuck about 100 bushels a day. By the end of the century, combines with eight-row heads could shuck and shell 100 bushels in less than 5 minutes!

Fruit combine

Soon just about anything could be harvested mechanically. Nuts are now gathered by machines that grab the trees and shake them, a method that also works for fruits such as cherries, oranges, lemons, and limes. Even tomatoes and grapes, which require delicate handling to avoid bruising, can be harvested mechanically, as can a diverse assortment of vegetables such as asparagus, radishes, cabbages, cucumbers, and peas.

Cotton harvester

Mechanical engineering ingenuity found solutions for even more problematic crops—the worst of which was probably cotton. In the long history of cotton's cultivation, no one had come up with a better way to harvest this scraggly tenacious plant than the labour-intensive process of plucking it by hand. It was in 1850 that a cotton harvesting device (called by some a cotton picker) was first patented by S. S. Rembert and J. Prescott, in Memphis, Tennessee U.S, but it was not until the 1940s that the machinery was widely used.

Mechanical cotton harvesters are of two types: strippers and pickers.

- *A stripper harvester* strips the entire plant of both open and unopened bolls, along with many leaves and stems. The cotton gin is then used to remove unwanted material.
- *Picker machines*, often called spindle-type harvesters, remove the cotton from open bolls and leave the bur on the plant. The spindles, which rotate on their axes at high speeds, are attached to a drum that also turns, causing the spindles to penetrate the plants. The cotton fibers are wrapped around the moistened spindles and then removed by a special device called a doffer; the cotton is then delivered to a large basket carried above the machine.

Cotton gin

The cotton gin, invented in 1794 by Eli Whitney, mechanized the post-harvest process of extracting the cotton fibers from the seedpod, or boll after it has been picked. Eli Whitney patented the cotton gin on March 14, 1794, but no real successful efforts at mechanizing the picking of cotton occurred until the 1930s.

In 1949, two brothers; John and Mack Rust of Texas developed the mechanical cotton picker. In that decade, demonstrated several different versions of a spindle picker, a device consisting of moistened rotating spindles that grabbed the cotton fibers from open bolls, leaving the rest of the plant intact; the fibers were then blown into hoppers. Spindle pickers produced cotton that was as clean as or cleaner than handpicked cotton; soon they replaced earlier stripper pickers, which stripped opened and unopened bolls alike, leaving a lot of trash in with the fibers. The Rust brothers' designs had one shortcoming: They could not be mass produced on an assembly line. Thus credit goes to International Harvester for developing the first commercially viable spindle picker in 1943, known affectionately as Old Red.

Tomato harvester

Mechanical tomato harvester development begun in 1950 by Blackwelder Machine Co., Rio Vista, CA, by Frederick L. Hill of Blackwelder (Ernest and Fred Blackwell), Coby Lorenzen, Jr., Steven J. Sluka, and vegetable crops specialist Gordie (Jack) C. Hanna of University of California, Davis, CA. The first successful machine was built in 1959 and became the dominant tomato harvester in USA. By 1968 tomatoes for processing were predominately harvested mechanically (see 1960/1961). In 1952, the first attempt in USA to mechanically harvest grapes was made by Winkler A. J. and Lloyd H. Lamouria in California. Shielded snapping rolls for corn harvesting were also developed by Charles Morrison, Deere & Co., Des Moines Works, Ankeny, IA same year.

Between 1960 and 1961, Successful tomato mechanical harvester was marketed by Blackwelder Manufacturing Co., Rio Vista, Ca, licensed by University of California, Davis in 1959 and patented in 1965, after considerable research and experimentation by Coby Lorenzen and Steven J. Sluka in cooperation with vegetable crops specialist Gordie (Jack) C. Hanna at the University of California, Davis. Blackwelder Manufacturing Co., Rio Vista, Ca. and Steven J. Sluka were the originators of the first successful tomato-vine separator used in the machine with contributions from Bill A. Stout of Michigan State University. By 1970 approximately 1500 units of the machine had been produced.

In 1996 Mechanical grape pruner, called Vinemaster, was developed by Roger Dellinger, and manufactured by the Valley Vine Machine Co., Richland, WA.

Mechanical weeders development

The earliest and the simplest of all technologies was manual weed control. Manual weed control started with farmers using their hands to uproot the weeds. The technology then advanced to hand tools, from using a stick to using a hand-hoe (Cloutier et al., 2007).

Manual weeding using human hands, provides a very effective weed control, but requires substantial human effort and energy. As agriculture became more mechanized, weeding tools were developed that were pulled by draft animals such as buffaloes and horses. As time progressed, these implements evolved and were adapted to tractors as the source of draft.

In mechanized agriculture, there are various types of mechanical weeding implements in the market that use three main techniques: burying weeds, cutting weeds and uprooting weeds. The burial of weeds is achieved through the action of tillage tools, and is usually done during land preparation. For cutting and uprooting weeds, there are two types of machinery available: inter-row weeders and intra-row weeders.

Figure 7-147: Inter-row rotary cultivator (Tornado, 2011).

Inter-row weeding is a weeding method that accomplishes between-planting row weeding, while intra-row does within-planting-row weeding. Mechanical inter-row weeders such as inter-row cultivators, rotary cultivators and basket weeders are available in the market (Cloutier *et al.*, 2007).

Rod weeder for fallow land farming was introduced in 1910 and rotary hoes (weeder) were produced commercially in 1912. The basket weeder is an implement consisting of several rolling rectangular-shaped wires, forming a round basket. There are also a wide range of mechanical intra-row weeders available (Mohd, 2012).

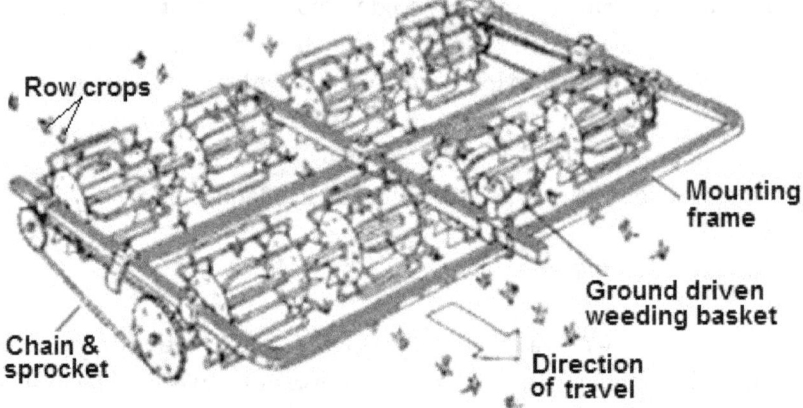

Figure 7-148: Basket weeder for inter-row weed control (Bowman, 1997)

Cloutier *et al.* (2007) and Weide *et al.* (2008) reported the usage of finger weeders and torsion weeders. A finger weeder is a simple mechanical intra-row weeder that uses two sets of truncated steel cone wheels with rubber spikes, or 'fingers' that are pointed horizontally at an angle toward the crop.

Figure 7-149: Finger weeder (Weide *et al.*, 2008)

Torsion weeders use flexible spring tines connected to a rigid frame and bent so that two short segments work close together and parallel to the work surface. Torsion weeder uses flexible coil spring tines to sweep the weeds.

Figure 7-150: Torsion weeder (Weide *et al.*, 2008)

One of the most promising technologies for intra-row weeding is the brush weeder. Cloutier *et al.*, (2007) reported that the brushes of the brush weeders are made of fiberglass and are flexible. These brushes can be vertically-rotated or horizontally rotated. Vertical-rotating brush weeder use hydraulics and require an operator to control the brushes. These weeders mainly uproot, but also bury and break weeds.

Figure 7-151: Vertical-rotating brush weeder (Melander, 1997)

The ECO-weeder is an intra-row mechanical weeder that is three-point hitch mounted and trails behind a tractor. It uses the tractor's power take-off (PTO) to drive a belt system that powers two discs with tines (Figure 7-152). This machine is quite similar to the brush weeder described above, but uses a mechanical drive and does not require any hydraulic power. ECO weeder uses rotating weeding mechanisms with tines.

Figure 7-152: ECO weeder (Hillside Cultivator Company, 2011).

The appearance of herbicides in the mid-20th century contributed to a decreased reliance on mechanical weeders (Cloutier et al., 2007). Gianessi and Reigner (2007) reported that during those years, labour became scarce and more expensive especially after World War II. Currently, however, it is becoming increasingly difficult to ignore the usage of herbicides in weed management because of its effectiveness to accomplish weed control and at the same time reduce yield loss (Mohd, 2012). However, renewed interests in mechanical weeding have grown due to environmental concerns, the growing demand for pesticide free produce and also the growth of herbicide-resistant weeds (Upadhyaya & Blackshaw, 2007).

Figure 7-153: A crop-row flame weeder using LPG gas (Physical Weeding, 2011)

There are also other types of non-chemical weed control methods such as flame weeding, pneumatic weeding, and laser weeding. These methods require other sources of energy to control weeds. The flame weeder, for example, requires propane gas to produce heat which elevates the temperature of the weed plants and either burns the weed biomass or causes weed plant cells to rupture and damage the plant structure (Figure 7-154).

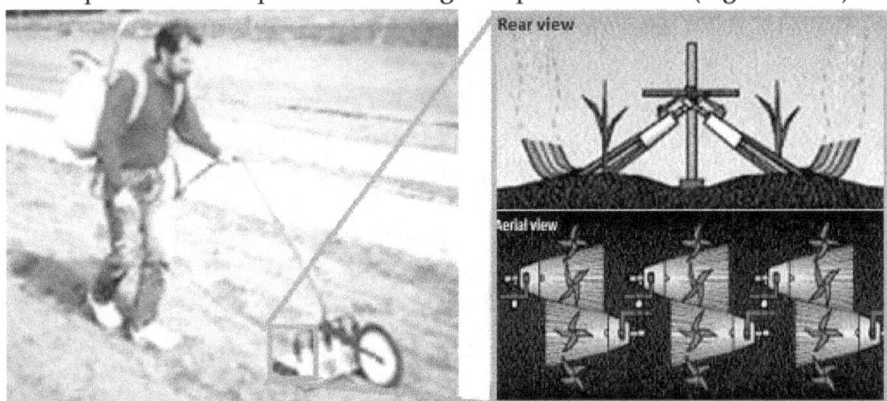

Figure 7-154: Flame weeder (inset: Rear and aerial views)

Pneumatic weeders require an air compressor, which injects compressed air into the soil to loosen and uproot small weeds (Figure 7-155; Bond et al., 2003). Both of these methods have substantial energy requirements. The flame weeder requires 28.2 to 131 liters of fuel per hectare (3 to 14 gallons of fuel per acre), depending on the intensity and coverage.

The pneumatic weeder uses substantial power, requiring a 60kW tractor to produce high air pressure to control weeds in well-anchored crops. This is twice the power required for conventional hoeing (Weide et al., 2008).

Figure 7-155: Pneumatic weeder uses air to blow out weeds (Weide et al., 2008)

Automated weeding technology

Advances in computers and sensors have contributed in the use of automation for agriculture machinery generally, and for weeding machines specifically. With automation, the weeding is operated electronically which reduces human intervention and optimizes the power provided by the machine. Automated machines also offer the possibility to determine and differentiate crop from weeds, and at the same time, remove the weeds with a precisely controlled device (Bakker, 2009). Automated weeder machine uses hydraulics to rotate semi-circle discs that are used for weed control.

Figure 7-156: Automated weeder machine (Tillett et al., 2008)

Astrand and Baerveldt (2002) developed an agricultural mobile robot using a perpendicular rotating weeding tool for weed control and two cameras – one near-infrared filter camera to locate crop row position and another colour camera to identify crop plants. Cloutier et al., (2007) reported on the "Sarl Radis" hoe developed in France. This automated weeder used light interception for crop detection, and a control system that controlled the lateral motion of a hoe relative to the crop row and around the crop plants. Sarl Radis intelligent weeder from France uses an automated hoe that moves in and out of the crop row.

Figure 7-157: Sarl Radis intelligent weeder (Cloutier et al., 2007)

Mowing machine

In 1822, mowing machine that used a horizontal cutting disc, for cutting hay and grass was invented and patented by an American, Jeremiah Bailey. Cylinder-cutting lawn mower developed and patented by Englishmen Edwin B. Budding (1796-1846), and John Ferrabee in 1830. 1844 marked remarkable improvements in the development of mowing machine by several patents issued including those of William F. Ketchum. Commercially manufactured gasoline-engine driven mower by Ransomes, Sims & Jeffries, Ltd. in England was introduced in 1902. In the 1930s, rotary-blade mowers were introduced.

Figure 7-158: Brush hog (Bush Hog)

In 1965 Hyaline mower conditioners for hay was manufactured by New Holland, Inc., New Holland, PA, based on work of H. G. McCarty, Lawrence Scrooge, and Elmer Del. The machine led the way for hay harvest methods and implements.

Figure 7-159: Mower attachment

Sprayers and spreaders

Commercial spraying machine for crops was introduced into agricultural production in 1880 and by 1884; a sprayer prototype was built by Frenchman Vermorel, based on idea put forth in 1781 by Father Rosier and by 1889, mechanical power-sprayers were commercially available. The manufacture of manure spreaders began in 1890, with one of which was the popular Flying Dutchman manufactured by the Moline Plough Co.

Attempt to control insects using airplanes was made in 1918, followed in 1921 by a specially equipped airplane (World War I Curtis JN6H plane by C. R. Neillie, Troy, OH.) for dusting crops, initially for cotton, and in 1924 by the first specially designed airplane for crop dusting. Air-blast type field and orchard sprayer was introduced in 1925

7.5 History of animal products processing

Milking machine

Development of a usable milking machine took several decades of trial and error, unlike the rapid development and acceptance of other dairy innovations, such as hygienic milking and processing, the Babcock test, and the centrifugal cream separator. In 1879, Anna Baldwin patented a milking machine that replaced hand milking - her milking machine was a vacuum device that connected to a hand pump. This is one of the earliest American patents; however, it was not a successful invention. Successful milking machines appeared around 1870.

Early cow milking machines

The great variety and number of early milking machines can be categorized into two groups, those that tried to emulate hand milking (mechanical pressure devices), and those that tried to emulate the sucking calf (vacuum devices). Proponents of both types of milkers turned out an endless variety of contraptions for over 50 years, until the modern pulsator made the suction method the clear winner.

The earliest devices for mechanical milking were tubes inserted in the teats to force open the sphincter muscle, thus allowing the milk to flow. Wooden tubes were used for this purpose, as well as feather quills. Skillfully made tubes of pure silver, gutta percha, ivory, and bone were marketed in the mid-19th century, and, in fact, a few were still being sold well into the 20th century. A novel milking tube illustrated in the *Scientific American* in 1875 used a slide valve at the bottom of each catheter to close off the opening. Catheter milking was blamed for various problems, such as spread of disease, weakened sphincter muscles causing continuous dribbling, and injury to the teats.

Cow milker patents

The earliest vacuum milkers used a large gutta percha cup, fitting over the entire udder, and connected to a hand pump. During the last half of the 19th century, over 100 milking devices were patented in the United States. Hodges and Brockenden secured an English patent for such a device in 1851. In America, Anna Baldwin patented such a milker, using a pitcher pump and bucket in her patent illustration. In 1859, S.W. Lowe, of Philadelphia, patented a cup fitted with a diaphragm with 4 holes for the teats. A hand cranked suction pump drew milk from all four teats at once. Such devices created a continuous suction on the udder, damaging the mammary tissue and frequently causing the cow to kick.

In 1859, John Kingman, of Dover, NH, patented a tin teat cup with elastic flange for use with a suction pump milker. The first successful use of teat cups with a vacuum milker is found in the 1860 patent of L.O. Colvin, perhaps America's most famous inventor of early milking machines. This lever operated suction device drew a great response from the agricultural press.

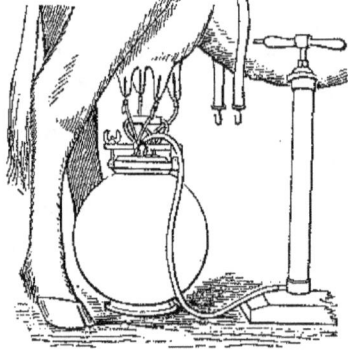

Figure 7-160: Cow milker machine

However, the Colvin milker still subjected the cow's teats to constant vacuum, causing blood to pool there. Colvin sold the English patent for this machine for $5000, and, at least 1500 machines were sold in England, according to an article in *The Agricultural Gazette*. In the U.S., Colvin was even more successful, and, continued to make improvements and acquire new patents.

In Scotland, William Murchland invented a very successful vacuum milker in 1889, which hung suspended under the cow. He was granted a U.S. patent in 1892. The Murchland milker, along with the famous "Thistle" milker, was extensively tested by the Highland and Agricultural Society of Scotland in 1898. Numerous other hand pumped suction milkers were devised in the next thirty years, with the foot operated Mehring machine being, perhaps, the ultimate in pre-pulsator suction milkers.

During the late 19th century, while many inventors were struggling with the problems of the constant suction milkers, others were working on a great variety of mechanical devices to simulate hand milking. Most of these devices incorporated rollers or fingers that intermittently pressed on the teat, often working from top to bottom. Some of these devices were simple; others were composed of hundreds of parts and worked by cranks. Such mechanical milkers were still being patented after the turn of the century, despite the arrival of the pulsator machines. Mechanical milkers could not compensate for the changing size of the cow's teats as milking progressed, and did not milk to completion. They also forced some milk back into the udder.

With so many inventors applying themselves to the task, the development of a satisfactory milker spanned over 50 years because of inherent risks on the cow during testing, or at least, her milk production. Farmers were understandably reluctant to offer their herd as guinea pigs, and this may have been the greatest obstacle to development of the milking machine.

7.6   History of farm transport development

Introduction

Farm transportation is as old as human existence because even the early man who was only a gatherer still had to convey himself to the centers of food collection. Early form of transportation was mainly on-farm as the major activities include the collection of water, crop gathering, animal hunting and related activities most of which were done within the neighborhood of the farmer and hence the distances covered were usually very short (Mijinyawa and Adetunji, 2005). History of the development of farm transport system dated back to the era when head carrying aids and hand carts were extensively used for crop packaging and transport.

The history and development of some of these carrying aids and vehicles whose history had changed the world forever, by adding a wheel to a hand operated vehicle, and it became possible for one man to do the work of two; not only that could he can now do the same work, he could move more swiftly and with less effort, had made its invention significant and the history of few of these inventions are discussed below.

Farm vehicles

There exist a wide range of low cost vehicles for moving farm goods, which can be categorized as:

1. Single carrying aids,
2. intermediate vehicles and
3. Advanced equipments.

Single carrying aid equipment

Single carrying aids for head, shoulder and back loading such as baskets bags, sacks etc. are significant in the history of farm vehicle development and as such its history worth mention.

Baskets: A basket is a container woven by hand fitted with either a lid, or left open which is traditionally constructed from stiff fibers, which can be made from a range of materials, including wood splints, runners, willows (in Britain), raffia and cane (elsewhere in low income countries) grown for that purpose in swampy land or river valleys. While most baskets are made from plant materials, other materials such as horsehair, baleen, or metal wire can be used. Baskets are generally cheap, light weight, easily repaired, durable and fairly strong. Most modern basketworks are exported and has been made on a frame by semi skilled workers. Twined baskets date back to 7000 BCE in Oasis America, while baskets made with interwoven techniques were common at 3000 BCE.

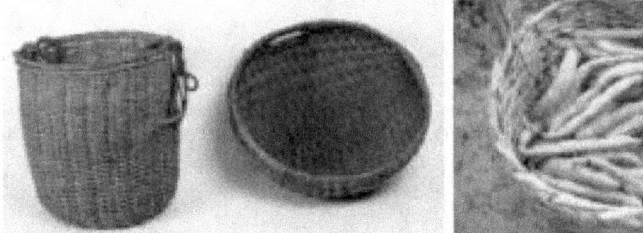

Figure 7-161: Traditional baskets

The history of the development of baskets date back thousands of years ago. Archaeological sites in the Middle East show that weaving techniques were used to make mats and possibly also baskets, circa 8000 BCE. Traditional basket making in Britain went into decline after the World War II, but were still in common use locally till date.

Figure 7-162: Load carrying aids for head

Sack bags: Donkeys were sometimes used to transport the corn to the threshing floor in the ancient Upper Egypt, but mostly it was carried by two men in a sack, fastened to a wooden frame and connected to five meter long carrying poles.

Figure 7-163: Traditional baskets

Wheel barrows: The invention of the wheel and the wheelbarrow had changed the world forever. A wheelbarrow is a small hand-propelled vehicle, usually with just one wheel, designed to be pushed and guided by a single person using two handles to the rear or by a sail to push the ancient wheelbarrow by wind. The term "wheelbarrow" is made of two words: "wheel" and "barrow." "Barrow" is a derivation of the Old English "bearwe" which was a device used for carrying loads.

Figure 7-164: Ancient wheelbarrow paintings from murals

The earliest descriptions and designs of wheelbarrows in the form of a one-wheel cart can be traced back from 2nd century Han Dynasty tomb murals and several other similar evidences between 118AD and 150AD of paintings of a man pushing a wheelbarrow.

There are evidences from some inventories of list building materials from the temple of Eleusis between 407 and 406 BC that the wheelbarrow may have existed in ancient Greece in the form of a one-wheel cart. Although evidence for the wheelbarrow in ancient Greek farming and mining is absent, it surmised that wheelbarrows were not uncommon on Greek construction sites for carrying moderately light loads.

Figure 7-165: Wheelbarrows

There are no documentary evidences of use of wheelbarrows by Roman builders although this may be due to scant surviving evidences while the present evidence does not indicate any use of wheelbarrows into medieval times. However, the Chinese historical Records of the Three Kingdoms (Sanguozhi) credited the invention of the wheelbarrow to Prime Minister Zhuge Chuko Liang (181-234 A.D.) of China. Liang was a general who used the wheelbarrows to transport supplies to injured soldiers. It was written that in 231 AD, Zhuge Liang developed the vehicle of the wooden ox and used it as a transport for military supplies in a campaign against Cao Wei. Further descriptions of the Liang's wheelbarrow design indicate a large single central wheel and axle around which a wooden frame was constructed in representation of an ox. The Chinese wheelbarrows had two wheels and required two men to propel and steer.

The wheelbarrow reappeared in Europe sometime between 1170 and 1250 AD (Lewis, 1994). Medieval wheelbarrows universally featured a wheel at or near the front (in contrast to their Chinese counterparts, which typically had a wheel in the center of the barrow), the arrangement now universally found on wheelbarrows. By the 13th century, the wheelbarrow proved useful in building construction, mining operations, and agriculture. However, going by surviving documents and illustrations the wheelbarrow remained a relative rarity until the 15th century and its applications had ever since remain relevant in Europe and the low income countries of the world.

In the 1970s, British inventor James Dyson introduced the Ballbarrow, an injection molded plastic wheelbarrow with a spherical ball on the front end instead of a wheel. Compared to a conventional design, the larger surface area of the ball made the wheelbarrow easier to use in soft soil, and more laterally stable with heavy loads on uneven ground. The Honda HPE60, an electric power-assisted wheelbarrow, was produced in 1998 (Honda worldwide, 1998)

Hand carts: The hand cart or 'barrow' is a small two-wheeled cart pushed by hand widely used in a variety of activities for the purposes of goods delivery. Hand carts were used as mobile shops, for example costermongers (selling fruit and vegetables).

Figure 7-166: Traders hand carts

The design of the hand cart was usually determined by the duties it was to perform. Hand carts were on the decrease from the mid 1930s, increasingly replaced by motorcycle combinations and small three and four wheeled vans, however, hand cart is still in common use today in the low income countries for refuse dumping, sawmill operations and in other typical operations.

Animal drawn carts: Early horse drawn carts were used for work delivery or for jobs such as hauling away of manure from stables. These carts are not fitted with a driving seat; the driver either walked alongside the horses head and holding onto the bridle or controlled by voice commands and the use of a long whip or rode one of the horses in the team. More specialized designs include the farmers or coal merchants 'tip cart' on which the body was mounted on hinges somewhere about its mid-point and could be tipped back to empty the load without un-hitching the horse. These tip carts are also known as tumbrils and were commonly called Scotch carts in Britain as the design originated in Scotland.

Figure 7-167: Tip cart

The tip cart has twin shafts, designed for a single horse secured between them, the farm cart has a rail to which may be attached either a pair of shafts (supplied as a U shaped set) or a sing pole, which could have a horse attached either side. The shafts tended to get in the way so when a cart with hinged shafts was 'parked' it was usual to get them out of the way. The use of dogs to pull carts was outlawed in 1837, but the term 'dog cart' remained in common use for a light weight horse drawn conveyance pulled by a small pony, were regularly seen into the 1940s.

Horse drawn carriage or wagon: Solutions to most of the early nineteenth century transportation problems and the improvements in road construction and maintenance had made horse drawn carriage a viable form of transport. The early horse drawn carriage was invented in Hungary in the 15th Century, but this early design had no steering as such and had to be dragged sideways by its team of horses to turn a corner. Development was slow but eventually the front axle was mounted on a turntable to provide steering. By the time the railways arrived a driver's seat was fitted to many passenger carrying vehicles (although it was much less common on goods vehicles).

Reins first appeared on these vehicles in the mid nineteenth century but carters still preferred to walk or sit on the front corner of the wagon. In town centers, the local councils preferred the driver to ride on the wagon and use reins to control the horse, this increased the speed of the vehicle somewhat and having the driver high up gave him a better view of the road.

Figure 7-168: 19th century traps, 1870s

In 1900 a law was passed requiring all wagons to be driven with reins but many carters continued to lead their horses. Up to the 1940s many people either rode a horse or used a horse drawn vehicle for personal transport. Once you added a solid roof and four wheels you were really in to two horse designs, which cost more to buy and run. The most common horse drawn vehicle would probably have been the small traders' delivery van, for which the designs varied greatly.

Figure 7-169: 'Hansom' cab and a Brougham type coach in use as a taxi, 1940s

Intermediate equipment (pedal driven vehicles)

Bicycles: In 1790, a Frenchman had popularized the use of an un-steerable foot-propelled two-wheeler mobile machine later known as bicycle. Thirty years later a German Baron took this idea and added both a sprung seat and steering. To allow the rider to use their hands for steering, a padded bar was added as a chest support.

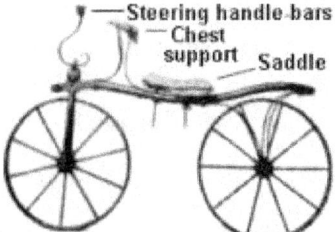

Figure 7-170: 'Hobby Horse', 1820

A Scottish blacksmith by the name Macmillan built himself a bike mechanically propelled by push rods (suitable chain-link had not been invented then) linked to a crank on the rear wheel in 1839; was not available so the bike was propelled. Macmillan took the machine for a spin but he overran and killed a child which caused him to abandon the research. In 1861, pedal was invented by a Frenchman called Pierre Michaux and within a year bikes with pedals mounted on the front wheel, called 'velocipedes', were being imported from Paris (although an Englishman by the name of Dennis Johnson had patented the idea as far back as 1818).

In 1869 the British produced the first machines with steel wheels and solid rubber tyres, by this time the name bicycle was in common use. The pedals were at this time mounted on the front wheel.

Figure 7-171: Velocipede bike, 1869

The speed of the bike could be increased by making the front wheel larger and during the 1870's front wheels became larger and in 1880 a British engineer called James Starley introduced the 'Penny Farthing'[21]. The large front wheel offered higher speeds and conferred a certain status on the owner, the design was popular and they became commonly known as the 'Ordinary'.

Figure 7-172: Penny farthing, 1880s

---

[21] The penny farthing or 'Big Bi' name referred to the relative sizes of the wheels (with a five foot diameter front wheel and a small trailing wheel) which resembled the British penny and the farthing or quarter penny coins. The legacy of the penny farthing is still with us today, if you look at the specification for bicycle gears they are quoted in 'inches', this figure is the equivalent diameter of the front wheel of a penny farthing type machine. Typical gearing on a standard bike today is quoted as '60 inches' and if you divide the gear figure of your bike by three you get the distance travelled in yards for one revolution of the pedals.

The big change in the history of 'bike' came at the later end of 1880's with the invention of chain driven 'safety bicycle' of the same general type as used today. The roller chain had been invented by Slater in 1864 and the first chain drive for bikes was developed by H. J. Lawson in 1879 but sales were slow until the revolutionary Rover Safety Cycle (designed by James Starley's nephew, John) appeared in 1885. Early models had a large front wheel and a slightly smaller rear wheel to retain the familiarity of the 'ordinary'. The design evolved quickly however and by 1890 bicycles had the familiar pattern tubular steel frame and pneumatic tyres on equal sized wheels.

Figure 7-173: Safety bicycle

The sprung saddle and ball bearings for the axles followed quickly, pneumatic tyres were introduced in 1887. In 1888 the bike was officially recognized in law as a 'carriage' and hence was formally allowed to use the roads, the same law required all bikes to be fitted with a bell which was to be continually rung whilst the bike was moving (that last bit of the law was finally repealed in Britain in 1930).

Figure 7-174: Bicycle transport equipment

Alex Moulton took a long look at the basic design of bicycles and came up with a new light weight variant in the later 1950s. His F-frame design which had no cross bar, and small wheels made it easier to get on and off the machine and to provide a comfort ride. The machine became instant success in 1962 and the Moulton style was widely imitated with varying degrees of success. The competition from larger firms caused Moulton to sell the rights to Raleigh in the early 1970s.

The designs in the 1970s with the long handlebar support were found to be problematic as (on some designs) the handlebars could come loose. In the early 1980s Mr. Moulton bought back the rights to his machine and produced a new even lighter version using an advanced 'space frame' design, called the AM which did well In the 1990s Pashley Cycles produced a low cost version under license. In the later 1980s a new type of bike appeared known as the 'mountain bike' and designed for people to ride off the roads and footpaths across open country.

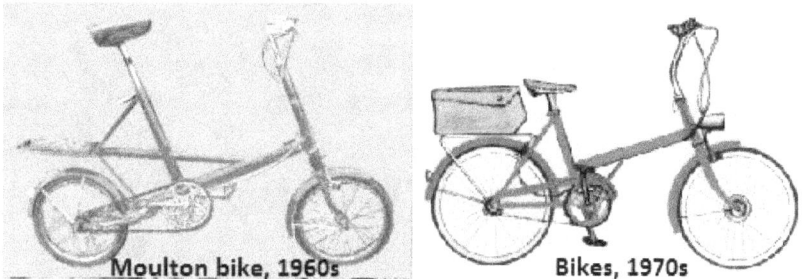

Figure 7-175: 20th century bicycle designs

Compared to other forms of transportation, the conventional bicycle is among the most efficient means of human locomotion. To travel one kilometer by bike requires approximately 5-15 watt-hours (w-h) of energy, while the same distance requires 15-20 w-h by foot, 30-40 w-h by train, and over 400 w-h in a singly occupied car (Justin, 2004).

Motorcycles and converted motorcycles: The power-assisted bicycle is an emerging form of transportation that attempts to merge the health and environmental benefits of a bicycle with the convenience of a motorized vehicle. Typical examples of power-assisted bicycle include motorcycles and tricycles.

Agricultural bikes are any motorbikes with two, three and four wheels used for agricultural works; on-farm or off-farm. Three and four wheelers are also known as 'all terrain vehicles' or ATVs.

Advanced equipment

Basic motorized vehicles with steam and gasoline powered engines were slowly but progressively developed through the centuries with applications in various fields of human endeavours including agriculture and transport. The dual-purpose agricultural transport equipment history equally gained attention at the turn of the 20th century.

All these vehicles have different advantages and disadvantages in terms of load bearing capacity, suitability for route conditions, running costs, speed range and capital cost which enable them to meet a broad spectrum of transport requirements. Many low cost forms of transport are used only in certain local areas and remain unknown even in other areas. After a century of the automobile, we can begin to assess the effects of long term transport by internal combustion. Nearly every aspect of our lives has developed around this technology. With the advent of new digital communications technologies, of the internet and beyond, which may eventually displace some of the functions of the automobile and replace our current problems with a new set that the next generation will be charged with solving.

Part 3

THE AGRARIAN REVOLUTIONISTS:
THEIR HISTORY & ENTERPRISE

Early tractor testing @ Nebraska

CHAPTER 8

AGRARIAN REVOLUTIONISTS
Their History

8.0 Introduction

Inventions and advances in farm power sources and machinery can be attributed to the product of many curious and ingenious agricultural inventors during the 19th and first half of the 20th centuries who either 'invented the wheels' or made incremental improvements to the work of their predecessors.

These individuals had mostly contributed to replacing ignorance with *knowledge*, savagery with *civilization*, disease with *health*, tyranny with *liberty*, poverty with *abundance*, and despair with *happiness*. Their story is one of intrigue, public condemnation, humiliation under pecuniary circumstances but definitely one of the greatest honours of being a useful instrument to changing human history that can never be taken from them, which far exceeds riches.

This chapter discusses the historical background of inventors who had made significant impacts related to agricultural production and development. Few of their history and inventions of significant are presented below.

8.1 The agrarian revolutionists: their lives and inventions

Jethro Tull (1674-1741)

Jethro Tull was born in Basildon, Berkshire, UK in 1674. He did not start out as an agricultural engineer. He studied law and graduated from Oxford University in 1699. Although he was admitted to the bar in the same year, he never practiced law; he later studied new techniques in agriculture during his travels across Europe.

Tull was far more interested in the farming methods employed on his inherited land in the southern part of England, which he called *Prosperous Farm* where he put into practice his study of agriculture.

On his return to Prosperous Farm in 1701, he developed a horse-drawn mechanical seed drill that would sow seed in uniform rows and cover up the seed in the rows. Up to that point, sowing seeds was done by hand by scattering seeds on the ground. Tull considered this method wasteful since many seeds did not take root.

The first prototype seed drill was built from the foot pedals of Jethro Tull's local church organ. The seed drill not only planted seeds at regular intervals but also planted them at the right depth and covered them with earth. Because the seed drill planted seeds in straight lines, a mechanical horse-drawn hoe, which Tull also invented, could be used to remove weeds from between the lines of crop plants.

Tull advocated the importance of pulverizing (crumbling) the soil so that air and moisture could reach the roots of the crop plants. His horse-drawn hoe was able to do this. He also emphasized the importance of manure and of tilling the soil during the growing season. At the time, Tull's ideas came under attack, mainly because they were new. His seed drill was not immediately popular in England, although it was quickly adopted by the New England colonists across the Atlantic.

In 1731, Tull wrote and published a book titled "The New Horse-houghing (hoeing) Husbandry" an essay on the principles of tillage and vegetation which he revised in 1733. Although his seed drill was improved in 1782 by adding gears to the distribution mechanism, the rotary mechanism of the drill provided the foundation for all future sowing technology.

Jethro Tull was widely regarded as the father of agricultural revolution and was part of a group of farmers who founded the Norfolk system, an early attempt to apply science to farming. Mr. Tull spent his lifetime and a fortune in developing this and other implements in the line of drills, horse-hoes, and cultivators, and died poor. His son died in prison for debt.

Andrew Meikle (1719 – 27 November 1811)

Andrew Meikle was an early brilliant Scottish mechanical engineer credited with the invention of the threshing machine, a device used to remove the outer husks from wheat grains. He also assisted Firbeck in the invention of the Rotherham plough. This was regarded as one of the key developments of the British Agricultural Revolution in the late 18th century. The invention was made around 1786, although some say he only improved on an earlier design.

Earlier, (c.1772), he also invented the windmill 'Spring sails', which replaced the simple canvas designs previously used with sails made from a series of shutters that could be operated by levers, allowing windmill sails to be quickly and safely controlled in the event of a storm. Meikle worked as a millwright at Houston Mill in East Linton, and inspired John Rennie to become a notable civil engineer.

He died at Houston Mill and was buried in East Linton's Prestonkirk Parish Church kirkyard, close to Rennie's father, George Rennie, who farmed the nearby Phantassie estate by the River Tyne. In 2011 he was one of seven inaugural inductees to the Scottish Engineering Hall of Fame.

Nicholas-Joseph Cugnot (1725-1804)

Cugnot was born at Void, in the Meuse province of France. As a young man, he joined the French army and while in the service in Germany and Belgium, he invented a new kind of rifle for use by French troops. He was also encouraged to work on a steam-powered gun-carriage. Cugnot was aware of the improvements in steam power developed by Thomas Savery, an English inventor, and Denis Papin, a French physicist.

Cugnot added further improvements, which employed steam power to move pistons without condensation, greatly improving engine efficiency. His engine consisted of two; 13-inch (33 cm), 1.75-cubic-foot (50-liter) pistons connected by a rocking beam which were synchronized so that when atmospheric pressure forced one piston up, high pressure steam forced the other piston down. The reciprocating motion was transferred to the axle, where it produced the rotary motion that turned the wheel. This arrangement is considered to be the first successful device for converting reciprocating motion into rotary motion.

Cugnot's first carriage had serious limitations. Its three-wheel design, with the boiler well out in front, was inherently unstable, and the whole heavy boiler-drive-wheel mechanism had to be turned to steer the carriage. It carried no reserves of water or fuel and required that the driver stop periodically to re-fire the furnace and add water to the boiler.

Despite these obvious drawbacks, the French Minister of War, the Duc de Choiseul, was encouraged by Cugnot's first demonstration, and he commissioned Cugnot to build a second larger, more powerful, and faster vehicle. This second carriage was completed by Cugnot in 1771 at a cost of 20,000 livres. Unfortunately, de Choiseul fell from power before this second vehicle could be fully tested, and his successor showed no interest in Cugnot's steam gun-carriage. It sat in a military shed for 30 years until it was moved to the Conservatoire National des Arts et Métiers in Paris, where it has remained on exhibit ever since.

Cugnot was granted a pension by the War Ministry in 1779, and he moved to Brussels. The French revolutionaries eliminated Cugnot's pension in 1789, but it was restored during the Consulate by Napoleon and continued until Cugnot's death on October 2, 1804, in Paris.

James Watt (1736-1819)

James Watt was a great Scottish engineer of the 18th century. He did not actually invent the steam engine. Instead he greatly improved it. A man named Thomas Savery invented the first primitive steam engine in 1698. A man named Newcomen started making steam engines to pump water from mines in 1712. However Watt is famous for inventing an improved version in 1769.

James Watt was born in Greenock on 19 January 1736. His father, also called James was a shipbuilder. As a boy Watt went to the local grammar school where he learned the classics and mathematics. However Watt also liked making models.

Eventually James decided to become a maker of mathematical instruments such as quadrants and compasses. In 1755 he went to London. However he did not stay there long. In 1757 Watt moved to Glasgow. In 1764 Watt mended a model of a Newcomen steam engine. In a Newcomen engine steam is admitted into a cylinder then condensed back into water. In 1765 Watt realized it would be more efficient to condense the steam in another chamber separate from the cylinder. However it was not until 1769 that Watt patented his knew idea, the separate condenser. Meanwhile in 1764 James Watt married a woman named Margaret Miller. The couple had 6 children but Margaret died after 9 years.

In 1766 Watt got a job as a land surveyor marking out land for canals. Then in 1774 James Watt moved to Birmingham. In 1775 he went into partnership with Matthew Boulton and began making steam engines. His steam engines were used for pumping water out of mines and gradually he became a wealthy man. Then in 1776 James Watt married Ann MacGregor. They had 2 children.

By 1780 the Industrial Revolution was beginning to transform life in Britain and Watt adapted his steam engine to provide a rotary motion so they could be used to power machines in the new factories. In 1781 he made the sun and planet gear to do this. In 1785 steam engines were used to power machines in cotton mills for the first time. Meanwhile in 1782 Watt invented another major improvement the double-acting steam engine. In 1788 Watt invented the fly ball governor to regulate the speed of steam engines and in 1790 he invented a pressure gauge.

In 1785 James Watt was elected a fellow of the Royal Society. James Watt retired from business in 1800. Watt died on 25 August 1819 and he was buried in Birmingham. Finally in 1882 a unit of electrical power was named the watt in his honour.

Richard Trevithick (1771-1833)

Richard Trevithick was an English inventor; the son of a mine engineer and as a child he would watch steam engines pump water out of the deep tin and copper mines. Trevithick was not simply an engineer - he was an inventor (his railway engines were running a decade before George Stephenson's) a visionary and an adventurer. His first contribution to steam development came when he used higher pressure steam, and got around the Watt patent by dispensing with the separate condenser.

In 1803 he built the "London Steam Carriage" which was basically a stage coach fitted with a steam engine. The following year, Trevithick built the world's first railway steam locomotive. Trevithick's locomotive had no name and was used at the Pennydarren ironworks in Wales, and pulled up to 10 wagons at speeds of around 5 mph (8 km/h). Although it worked, it was not financially successful as it was too heavy for, and kept breaking the rails (which were designed for animal propelled trains).

In September 1825, Stephenson completed the first locomotive for the new railroad; named at first Active, it was soon renamed Locomotion. As the Liverpool & Manchester approached completion in 1829, the directors of that company arranged for a competition to decide who would build the locomotives for the new railway. Stephenson's entry was The Rocket, and its impressive performance in winning the contest made it arguably the most famous machine in the world. Stephenson's son, Robert Stephenson, was also a noted locomotive engineer, and was heavily involved in the creation of many of his father's engines from Locomotion onwards.

Richard Trevithick referred to as the Father of the locomotive engine by Jeffrey Ezell was credited with many inventions over his lifetime, including the steam carriage, the steam barge, the portable agricultural engine, and the screw propeller. He was a pioneer of the Industrial Revolution and undoubtedly one of the greatest engineers to have ever lived. But he still ended his life in poverty wondering how it all went wrong. Trevithick was quoted as writing this note to Davies Gilbert in what could best be described as frustration:

I have been branded with folly and madness for attempting what the world calls impossibilities, and even from the great engineer, the late Mr. James Watt, who said to an eminent scientific character still living (in my time), that I deserved hanging for bringing into use the high-pressure engine. This so far has been my reward from the public; but should this be all, I shall be satisfied by the great secret pleasure and laudable pride that I feel in my own breast from having been the instrument of bringing forward and maturing new principles and new arrangements of boundless value to my country. However much I may be straitened in pecuniary circumstances, the great honour of being a useful subject can never be taken from me, which to me far exceeds riches.

George Stephenson (1781- 1848)

George Stephenson was born on June 9, 1781, in the coal mining village of Wylam, England. His father, Robert Stephenson, was a poor, hard working man that supported his family entirely from wages of twelve shillings a week. George Stephenson's first job was to watch over a few cows owned by a neighbour which were allowed to feed along the road; George was paid two cents a day to keep the cows out of the way of the coal-wagons; and also, to close the gates after the day's work of the wagons was over.

George Stephenson's next job was at the mines as a picker, a fireman, plugman, brakeman, and engineer. However, in his spare time George loved to tinker with any engine or piece of mining equipment that fell into his hands. He became skilled at adjusting and even repairing the engines found in the mining pumps, even though at that time he could not read or write. As a young adult, George paid for and attended night school where he learned to read, write, and do arithmetic. In 1804, George Stephenson walked on foot to Scotland to take a job working in a coal mine that used one of James Watt's steam engines, the best steam engines of the day.

In 1807, George Stephenson considered migrating to America; but he was too poor to pay for the passage. He began work nights repairing shoes, clocks, and watches, making extra money that he would spend on his inventing projects.

In 1813, George Stephenson became aware that William Hedley and Timothy Hackworth were designing a locomotive for the Wylam coal mine. So at the age of twenty, George Stephenson began the construction of his first locomotive. It should be noted that at this time in history, every part of the engine had to be made by hand, and hammered into shape just like a horseshoe. John Thorswall, a coal mine blacksmith, was George Stephenson's main assistant. After ten months' labour, George Stephenson's locomotive "Blucher" was completed and tested on the Cillingwood Railway on July 25, 1814.

The track was an uphill trek of four hundred and fifty feet. George Stephenson's engine hauled eight loaded coal wagons weighing thirty tons, at about four miles an hour. This was the first steam-engine powered locomotive to run on a railroad and it was the most successful working steam engine that had ever been constructed up to this period, this encouraged the inventor to make further experiments. In all, Stephenson built sixteen different engines.

George Stephenson built the world's first public railways: the Stockton and Darlington railway in 1825 and the Liverpool-Manchester railway in 1830. Stephenson was the chief engineer for several of the railways. In 1815, George Stephenson invented a new safety lamp that would not explode when used around the flammable gasses found in the coal mines.

Also in 1815, George Stephenson and Ralph Dodds patented an improved method of driving (turning) locomotive wheels using pins attached to the spokes to act as cranks. The driving rod was connected to the pin using a ball and socket joint, previously gear wheels had been used.

Stephenson and William Losh, who owned an ironworks in Newcastle, patented a method of making cast iron rails. In 1829, George Stephenson and his son Robert invented a multi-tubular boiler for the now-famous locomotive "Rocket".

Eli Whitney (1793-)

In colonial times, cotton cloth was more expensive than linen or wool because of the extreme difficulty of separating seed from the clinging fibers. One man could pick the seeds from only about 1 pound of cotton fiber per day.

In 1793, Eli Whitney built a machine consisting of a row of close-set wheels with saw-like teeth around their perimeters. The wheels protruded through narrow slits between metal bars into a hopper filled with cotton bolls. As the wheels revolved, the teeth caught the cotton fibers and pulled them through the slits, which were too narrow for the seeds to pass, thus separating the two. Whitney patented the cotton gin on March 14, 1794.

Whitney's cotton gin allowed 1,000 pounds of cotton to be cleaned in the time it took one man to do 5 pounds by hand. As a result, the price of cotton cloth plummeted, the cotton plantation culture of the South was established and the use of slave labor in growing cotton became entrenched.

Cyrus Hall McCormick Sr. (1809-1884)

Cyrus H. McCormick of Virginia was more than a mechanic; he was a man of vision; and he had the enthusiasm of a crusader and superb genius for business organization and advertisement. Grain harvesting machines first appeared in Great Britain in about 1800, and in the U.S. a decade or two later, but most failed.

In 1831, Cyrus H. McCormick developed the first commercially successful reaper, a horse-drawn machine that harvested wheat. Though Cyrus offered reapers for sale in 1834, he sold none in that year and for six years afterwards. He sold two in 1840, seven in 1842, and fifty in 1844. The reaper did not work well in the hills of Virginia, and farmers hesitated to buy anything that needed the attention of a skilled mechanic. McCormick's machine became the more popular one; today he is credited with inventing the reaper.

However, things changed after Cyrus McCormick made a trip through the Mid West. In the rolling prairies, with mile after mile of rich soil with few trees or stones, McCormick saw his chance. At this time, Obed Hussey had moved east. Cyrus McCormick did the opposite; he moved west, to Chicago, in 1847. Chicago was then a town of hardly ten thousand; Cyrus McCormick built a factory there and began the mass manufacture of his reaper, and manufactured five hundred machines in time for the harvest of 1848.

He was not daunted by the Government's refusal in 1848 to renew his original patent. He successfully decided to make profits as a manufacturer rather than accept royalties as an inventor. He formulated an elaborate business system. His machines were to be sold at a fixed price, payable in installments if desired, with a guarantee of satisfaction. He set up a system of agencies to give instruction or to supply spare parts. Advertising was done chiefly by exhibitions and contests at fairs and other public gatherings. Those early machines still required the sheaves to be bound by hand, but in 1857 the Marsh brothers equipped a reaper with moving canvases that carried the grain to a platform where it was tied into bundles by a worker riding on the machine.

The next innovation, patented in 1858, was a self-raking reaper with an endless canvas belt that delivered the cut grain to two men who riding on the end of the platform, bundled it. Meanwhile, Cyrus McCormick had moved to Chicago, built a reaper factory, and founded what eventually became the International Harvester Company. In 1872 he produced a reaper which automatically bound the bundles with wire. In 1880, he came out with a binder which, using a magical knotting device (invented by John F. Appleby a Wisconsin pastor) bound the handles with twine.

After the death of Cyrus, Sr., in 1884, Cyrus, Jr., renovated the marketing organization and focused manufacturing capabilities on producing a durable machine with the least complexity, largely limiting the company to a single model of each principal machine -- reaper, self-rake, mower, binder- to give maximum production runs. By the 1890s McCormick not only commanded a premium price -- often 10 percent or more above any rival- but built sturdy machines at lower cost than any competitor.

Obed Hussey (1792-1860)

Obed Hussey of Cincinnati was an avid inventor who rarely shied away from a challenge. He was especially intrigued by the development of new agricultural machines. Born in Maine, Hussey moved to Nantucket Island, Massachusetts, when he was very young. He probably worked as a sailor for a while. By the time he was thirty, he was in Cincinnati, Ohio, and working on his inventions.

In 1830 someone suggested to Hussey that he invent a machine for cutting grain. He had already invented, among other things, a corn-husking machine, a corn grinder, and a sugarcane press. He went to work on this latest challenge and travelled to Baltimore, Maryland, the following year to conduct his reaper experiments at the farm implement factory owned by Richard B. Chenaworth.

In 1832, Hussey returned to Cincinnati and built his first successful reaper. It was publicly demonstrated in July 1833 to the Hamilton County, Ohio, Agricultural Society. Hussey's machine consisted of a reciprocating saw tooth bar driven by gears turned by the main drive. The machine was horse-drawn. As the blade moved forward, the stalks of grain fell onto a platform where they were manually bound. This machine sold well enough between 1834 and 1838 in Illinois, New York, Pennsylvania, and Maryland that in 1838 Hussey opened a large factory in Baltimore.

Even as Hussey was developing his first reaper, Cyrus McCormick was working on one of his own at his father's forge in Virginia. McCormick's reaper was patented in 1834, and the two inventors entered into direct competition. The independent inventions of Obed Hussey in 1834 and that of Cyrus Hall McCormick Sr. brought about an end to tedious handiwork and encouraged the invention and manufacture of other labour-saving farm implements and machinery. Hussey was an excellent mechanic. He patented several improvements to his machine and received high praise for the efficiency of the work.

Both Hussey and McCormick built improved versions of their machines in an effort to stay ahead of each other, and both exhibited their machines at the Crystal Palace industrial exhibition in London in 1851. In the years following the exhibition, McCormick gradually eroded Hussey's standing in the reaper industry, shrewdly acquiring the patents of others. McCormick saw his chance to become the greatest figure in the world of agricultural machinery when he made a trip through the Mid West.

In the rolling prairies, with mile after mile of rich soil with few trees or stones, he built a factory and began the mass manufacture of his reaper. Hussey had intolerance for other people and obstinately preferred to improve his machines himself. The difference between the two men gave the competitive edge to McCormick, and he was soon outstripped in the race because he was weak in the essential qualities which made Cyrus McCormick the greatest figure in the world of agricultural machinery and in 1858, Hussey sold out. Obed Hussey had moved east, to Baltimore. He went to work once again, however, this time on a steam-powered plough. His work was cut short in 1860 when he died after falling under a moving train in New England.

Leander James McCormick (1819-1900)

Leander James McCormick (1819-1900) was an American inventor, manufacturer, philanthropist, and businessman. He was born February 8, 1819 in Rockbridge County, Virginia on February 8, 1819 to Robert McCormick, Jr. (1780-1846) and Mary Ann "Polly" Hall (1780-1853). His elder brother Robert Hall McCormick died as a teenager and his younger brother John Prestly McCormick also died as a young adult, but Leander survived.

His father invented agricultural machines including the mechanical reaper, for which Leander's eldest brother Cyrus McCormick Sr. received the patent in 1834. Along with his elder brothers Cyrus and William, they were regarded as one of the fathers of modern agriculture due to their role in the development of the McCormick Reaper and what later became the International Harvester Company. Leander eventually developed multiple improvements to the reaper and received patents for two of them, with the remainder being patented by his brother Cyrus.

In 1847 Leander helped Cyrus set up a factory in Cincinnati, Ohio that produced 100 machines. In fall 1848 he moved to Chicago with his wife and infant son to join Cyrus in setting up an even larger factory. Another older brother William Sanderson McCormick joined in 1850 in a business in run by Cyrus to manufacture reapers and sell them across the mid western United States. They created what eventually became the McCormick Harvesting Machine Company, with Leander taking over management of the manufacturing department, which he controlled for the next 30 years. By 1870, the McCormicks were one of the wealthiest families in the United States.

In 1871, the Great Chicago Fire destroyed much of the Reaper Works and other buildings, as well as the Leander McCormick family residence at the corner of Rush Street and Ohio Street. Leander quickly rebuilt and recovered. By 1879, the business had fully recovered and was merged into a corporation. Leander stayed active in the management of the business until 1889 when he retired. After retiring from the business, McCormick then invested heavily in real estate in downtown Chicago and Lake Forest, Illinois. In 1885 he donated one of the world's largest telescopes to the University of Virginia. He died on February 20, 1900 at the Virginia Hotel.

Nikolaus August Otto (1832-1891)

Nikolaus Otto (1832-1891) born at Holzhausen, Hesse-Nassau,, the son of a farmer, was a German inventor who created the four-stroke internal-combustion engine, the first engine to burn fuel directly in a piston chamber. Otto left school at 16, worked as a businessman in Frankfurt am Main and in Cologne in his early years. After relocating to Cologne where he became fascinated by the gas engines developed by Frenchman, Jean-Joseph-Étienne Lenoir displayed for the first time in 1859.

He quit his office job, starting out by seeking to improve on the existing design of Étienne Lenoir, in 1861; Otto built an experimental engine based on Lenoir's design. Prior to his invention, all engines were of an external-combustion design, and fuel was burned in a separate compartment. Three years later, Otto met another industrialist engineer Eugen Langen in 1864. The technically trained Langen recognized the potential of Otto's development, and one month after the meeting, founded the engine factory, NA Otto & Cie.

In 1867, Otto and Langen announced their first production engine--a noisy two-stroke engine that improved upon the Lenoir design by compressing the gas before it was ignited. At the 1867 Paris World Exhibition, their improved engine was awarded the Grand Prize.

Otto later turned his attention to the 4-stroke cycle (as described in a pamphlet by Alphonse Beau de Rochas in 1862). This was largely due to the efforts of Franz Rings and Herman Schumm, brought into the company by Gottlieb Daimler. It is this engine (the Otto Silent Engine), and not the Otto & Langen engine, to which the Otto cycle refers. This was the first commercially successful engine to use in-cylinder compression (as patented by William Barnett in 1838). The Rings-Schumm engine appeared in autumn 1876 and was immediately successful.

The four-stroke cycle engine, or Otto cycle engine, is the most widely used type of engine in automobiles, large boats and light aircraft. Though the concept of four strokes, with the vital compression of the mixture before ignition, had been invented and patented in 1861 by Alphonse Beau de Rochas, Otto was the first to make it practical.

Otto also made the internal combustion engine mobile by using a low-voltage magneto ignition which allowed the engine to use liquid fuel. His compact, powerful and mobile engine helps to empower millions of people.

Despite Otto's commercial success, he encountered legal problems over the patent of his four-stroke engine when several others challenged his right to the invention. Although the courts invalidated Otto's patent in 1886 due to the claims of Beau de Rochas and others, the Otto-cycle engine's superiority assured its continued success. It is the Otto-cycle engine that made possible automobiles, motorcycles, powerboats, and aircraft, all of which require a small, light, and efficient engine. Nikolaus Otto died in Cologne on January 26, 1891.

John Deere (1804-1886)

John Deere was born in February 1804 in Rutland, Vermont, on February 7, 1804. He was an Illinois blacksmith and manufacturer. His father left for England and disappeared in 1808, and, subsequently, Deere was raised by his mother. He was educated in the public school system and began his storied industrial career as a blacksmith's apprentice at age 17, setting up his first smithy trade just four years later. He spent the next 12 years keeping busy with his trade in various towns around Vermont.

As a blacksmith, he was making the same repairs to ploughs again and again, and realized that the wood and cast-iron plough used in the eastern United States—designed for its light, sandy soil—was not up to the task of breaking through the thick, heavy soils of prairieland. Early in his career, Deere determined that the wood and cast-iron plough in use at the time was ill suited to the challenges presented by prairie soil, so after some experimentation. In 1837, on his own, he invented the first self-polishing cast steel plough that greatly assisted the Great Plains farmers and sold his first one in 1838, and 2 others before the end of the year.

Facing a tough business environment, in 1837, a 33-year-old Deere packed up and headed west, eventually settling in Grand Detour, Illinois. There, he set up another blacksmith shop. The following year, he sent for his wife, Demarius Lamb, and their five children. The large ploughs made for cutting the tough prairie ground were called "grasshopper ploughs." The plough was made of wrought iron and had a steel share that could cut through sticky soil without clogging.

John Deere became a millionaire selling his steel ploughs. During World War I with the rise of farm prices and the demand for dependable mechanical farm power, the concept of the tractor became so popular that in a matter of months many tractor manufacturers sprung up.

Since Deere was looking for an established farm tractor to round out its line, it was decided that here was an organization with many years experience – a company that knew what farmers wanted – what it took to build a good tractor.

John Deere was active in the community of Moline, Illinois, throughout his life, even serving as the city's mayor for two years. He died on May 17, 1886, at his home in Moline.

Meinrad Rumely (1823-1904)

Meinrad Rumely, was a German immigrant who came to the United States in 1848 at the age of 25 from Havre, France. On arrival, it took him 64 days to reach New York. His brothers had arrived the U.S. two years earlier in 1846. Meinrad passed through Castle Garden and it was there all of his money was taken from his trunk. Fortunately, he found a job in a machine shop which gave him enough money to go further west. He came to La Porte, Indiana where there were already several people in business who had come from Germany and who urged him to start a machine shop and foundry.

Like many similar stories during this time, the flow of technology followed the immigrants. Meinrad settled in La Porte because there was a new railroad shop and yard being built to support the main line between New York and Chicago. In 1853, the two brothers, Meinrad and John, started their small shop in La Porte (this shop later moved to Elkhart in 1870) for the manufacture of corn shelters and horse-powered machines and the casting of irons for the railroad shops. That was the beginning of the Rumely Companies.

Meinrad Rumely had learned the millwright trade in France before he immigrated to the United States in 1848. He had the ability to visualize complex machines and their mechanisms. He could organize and manage people, materials, and processes. These qualities made the company quickly grew to 15-20 employees and were pouring 1 tone of castings a day. It produced the separator (threshing machine) in 1857, and two years later, won first prize for threshing at the Illinois State Fair in Chicago. The brothers soon began to produce stationary steam engines. By 1869, they were the town's largest employer, employees totaled 35 and sales were $50,000 a year.

In 1872, the Rumely Company developed a portable steam engine that could be horse-drawn from one farmyard to another and linked to a thresher with a driving belt. A decade later, it introduced a steam traction engine that pulled the threshing machine and water wagon. (The brothers had won eight first prizes in threshing by then.) In 1886 a new straw-burning engine was introduced. A new foundry opened in 1890. A year later, a self-feeder was added to the line. By 1892, a new boiler and blacksmith shops were added and also a clover hulling machine was added in 1901.

At Meinrad Rumely's death in 1904, the company had 300 employees and had been granted at least five U.S. patents related to agricultural equipment.

William Deering (1826–1913)

William Deering was born on April 25, 1826, in South Paris, Maine, a small town about 40 miles north of Portland. His family owned a wool mill in South Paris, where young William apprenticed earning $18 per month. William Deering is not an inventor but more of an entrepreneur who had different origins and strategy. Even though he was one of the most-instrumental figures in American agricultural equipment development before and after the turn of the 20th century, William Deering's contributions to the industry remain an enigma to some collectors. He never designed a single piece of machinery, but Deering managed to turn other people's ideas and inventions into an empire that eventually became International Harvester Co.

While at the wool mill, as a resourceful man, he rose through the management ranks until 1850, when he traveled west to invest in land in Illinois and Iowa. Like many pioneers of his day, the frontier didn't suit him. As a result, in 1856 Deering returned to Maine, where he opened a dry goods store in Portland called Deering Milliken & Co. Around 1870 Deering had been a successful and wealthy dry goods merchant in Maine selling uniforms to the government during the civil war before leaving the business and moving to Chicago to partner with Elijah Gammon, providing $40,000 in funding for the production of a horse-drawn grain harvester developed by two brothers; William Marsh and Charles Marsh.

Deering conducted several experiments in 1880 and determined that the ideal binder twine would be made of manila, spun to 700 feet per pound. He induced Edwin Fitler of Philadelphia to produce a hard-twist manila twine, and made the binder an overwhelming success. Deering was also responsible for building a modern twine factory to supply farmers with sufficient length and quality of twine to work with the binders, a move followed by most competitors.

By 1885 the twine-binder made Deering probably second only to McCormick in harvester production - and may have led him past the venerable leader, briefly, in 1887 and 1888. But Deering never developed a marketing organization comparable with McCormick's; by 1890 it had fallen back into the second position it would occupy until creation of International Harvester Co.

Deering preferred to compete on the basis of innovation in design, integration of production, and, to a degree, on price. Thus Deering was among the leaders in developing all steel binders, in applying ball and roller bearings, and in developing light binders. Deering was the first harvester manufacturer to have its own twine mill and aggressively brought production of most components of its machines into its own factory. It anticipated McCormick in expanding into new lines, particularly corn binders and hay rakes, acquiring a Canadian factory and buying its own steel mill.

As early as 1896 the Deerings had begun considering making and rolling their own steel; by 1897 the volume they required made a mill both more attractive for cutting cost and more important to protect themselves from suppliers unwilling to handle their orders - only four or five mills had sufficient capacity. Only William Deering's opposition delayed acquisition until 1900 when dramatic cost increases drove home the vulnerability of their position. By 1902, with this aggressive strategy, Deering stood far ahead of any competitor save McCormick itself, though it was probably no more than three-fourths the size of the leader.

William Deering financially supported several institutions of Chicago, the Northwestern University, the Garrett-Evangelical Theological Seminary, and the Wesley Hospital among them. He gave Northwestern over $1 million over the years, and served on the university's board for 38 years, including 10 years (1895–1905) as president of the board; he declined an offer to rename the school *Deering University*.

After Deering retired in 1901 he spent a large part of each year at his winter home in Coconut Grove, Florida. He died on December 9, 1913 in Coconut Grove with his two sons in attendance. He was the father of Charles Deering (1852–1927) and James Deering (1859–1925).

William Wallace Marsh (April 15, 1835 – May 12, 1918) & Charles Wallace Marsh (March 22, 1835 –)

William Wallace Marsh was born near Trenton, Ontario on April 15, 1835 to Samuel and Tamar Marsh. Samuel Marsh was a Patriote in the Rebellions of 1837, captured and jailed for five months. William studied at Victoria College, Cobourg for three years and decided to move with his father to a 110-acre (45 ha) farm in Clinton Township, DeKalb County, Illinois near Shabbona Grove in 1849 at his return from jail.

The farm proved difficult for the family to manage. In 1857, William Wallace Marsh began to experiment with grain bundling methods. Marsh discovered he could bind grain while another was being cut if the cutting apparatus could move. His brother, Charles W. born on March 22, 1834, in Northumberland, County, Ontario, Canada, joined the effort the next day. For the next few years, the brothers worked together to perfect their harvester.

Marsh Brothers was founded in 1858. William Marsh was the inventor of the company while his brother Charles handled the financial aspects. The Marsh brothers patented their device in 1858, but farmers were slow to accept the new machine, despite its usefulness. In 1859, the Marsh brothers unveiled the completed product during a reaping contest north of DeKalb.

The success of the Marsh's design was owed to the fact that during demonstration anyone could use the implement. The system involved movable canvases coupled with a platform and table that allowed one or two men to ride a reaper while tying the cut grain into bundles. Before that innovation, cut grain was raked off the reaper in gavels, (or straight piles) and then men who followed the machine, gathered each gavel and tied it into bundles.

William Marsh went to Piano, Ill., in the winter of 1860, where he formed a company with a prominent settler, Lewis Steward, who invested in the company to build the Marsh harvester. Originally capable of producing 100 harvesters in a year, Steward's capital allowed the company to expand to 10,000 per year. The Marsh brothers withdrew from the Plano group at about 1869, and started their own plant in Sycamore, Ill. Elijah H. Gammon, a relatively wealthy man and former Methodist preacher, were soon involved with the firm in the 1870s. That was great for the Marshes, because the reaper business required a lot of capital to operate.

In 1870, their business got an added financial boost and a new partner in the person of William Deering two years later. Marsh married M. J. Smith on January 8, 1871. William Marsh was elected alderman of Sycamore is 1873, a position he held for decades. William Deering purchased the company in October 1875 and moved most of its operations to North Chicago. It merged with the McCormick Harvesting Machine Company in 1902 to form International Harvester.

Steward & Company moved to Sycamore in 1873. Marsh operated the Sycamore division until his retirement in 1906. He died there on May 12, 1918. His 1873 house in Sycamore was listed on the National Register of Historic Places by the National Park Service on December 22, 1978.

John Francis Appleby (1840-1917)

John Francis Appleby (1840–1917) was an American inventor who developed a knotting device to bind grain bundles with twine, which became the foundation for all farm grain binding machinery and was used extensively by all the major manufacturers of large grain harvesting machines in the late 19th and early 20th centuries. John Francis Appleby was born in Westmoreland, New York in 1840. In 1844 his extended family of 17 arrived by boat to Milwaukee, Wisconsin.

From 1862 to 1865 He served with the 23rd Wisconsin Volunteer Infantry Regiment in the American Civil War. During the war, Appleby invented and patented a manual magazine feed breech loading needle gun. When the US government rejected the idea, Appleby sold the patent for $500. The weapon was later used extensively by the Prussian Army. When he was just 18, Appleby invented and demonstrated his first wire knotter in 1867. The knotting device was to become the foundation for all farm binding machinery, but no one was interested in the idea at the time. Appleby's knotting device was a major landmark in the mechanization of agriculture and aided the development of the western wheat fields of the United States.

He was unable to gain any financial backing for it because of lack of support from farmers for the use of wire binding because bits of wire got into the grain and ended up inside livestock and flour with disastrous results. By 1878, Appleby had developed a successful twine binder, which he patented. Twine binders did not cut into the wheat or, like wire binders, kill cattle that happened to eat a strand.

He licensed the twine binder mechanism to the Gammon and William Deering adopted the twine-tying mechanism for his popular Deering harvesters and in about 1881, McCormick did adopt it as well. Appleby's design soon became the standard grain binding device used on machines manufactured by Cyrus McCormick's Harvesting Machine Company, Champion Machine Works, and the D.M. Osborne Co.

In 1881, Appleby sold his grain binder machine patent interests to Champion Machine. He continued to work on various inventions, eventually patenting a horse-drawn cotton harvesting machine in 1905. He died in Chicago in November 1917.

John Froelich (1849-1933)

John Froelich was an American inventor 24th November, 1849 and lived in Waterloo, Iowa. John Froelich attended school in Galena, Illinois and at the College of Iowa. At the end of the 19th century, Froelich operated a grain elevator and mobile threshing service: Every year at harvest time, he dragged a crew of hired hands and a heavy steam-powered thresher through Iowa and the Dakotas, threshing farmers' crops for a fee. His machine was bulky, hard to transport and expensive to use, and it was also dangerous: One spark from the boiler on a windy day could set the whole prairie afire. So, in 1890, Froelich decided to try something new: Instead of that cumbersome, hazardous steam engine, he and his blacksmith mounted a one-cylinder gasoline engine on his steam engine's running gear and set off for a nearby field to see if it worked.

It did: In 1892, he developed the first stable gasoline/petrol-powered tractor with forward and reverse gears. Froelich's tractor chugged along safely at three miles per hour. But the real test came when Froelich and his team took their new machine out on their annual threshing tour, and it was a success there, too: Using just 26 gallons of gas, they threshed more than a thousand bushels of grain every day (72,000 bushels in all). What's more, they did it without starting a single fire.

In 1894, Froelich and eight investors formed the Waterloo Gasoline Traction Engine Company to manufacture the "Froelich Tractor, and made John the president". Unfortunately, efforts to sell the practical gasoline-powered tractor failed. They built four prototype tractors and sold two (though both were soon returned). The company then decided to manufacture stationary gas engines to provide income while tractor experiments continued. Froelich was more interested in farming equipment than engines more generally, however, and he left the company in 1895.

Waterloo kept working on its tractor designs, but between 1896 and 1914 it sold just 20 tractors in all. In 1914, the company introduced its first Waterloo Boy Model "R" single-speed tractor, which sold very well: 118 in 1914 alone. The next year, its two-speed Model "N" was even more successful. John Froelich will always be remembered and the village of Froelich, Iowa boasts the name "Tractor Town U.S.A."

Charles Walter Hart (1872-1937) and Charles H Parr (1868-1941)

Charles W. Hart and Charles H. Parr met as mechanical engineering students in 1892 at the University of Wisconsin in Madison, Wisconsin. Both had grown up in agricultural areas with labour-intensive family farms. Naturally, they both had interests in developing labour-saving farm equipment, and they became close friends.

Their mutual interest in the development of the internal combustion engine led them to a joint, extra credit project to produce an internal combustion engine and graduated with honors from college in June 1896. In fact, even before graduation, they started manufacturing and selling engines for farm use from their factory shop in Madison. They produced five engines.

While still in college, they opened a small machine shop in Madison, Wisconsin, where they repaired damaged farm implements and began experimenting with the internal combustion engine. One of the requirements for their degrees was a senior thesis consisting of an original investigation that would contribute to the knowledge of engineering or the design and construction of a device useful to industry. Hart and Parr worked together on their thesis developing a design for a new type internal combustion engine. In conjunction with their thesis, they built in their shop a working model of the new internal combustion engine type that their thesis described.

After graduation, they sought and secured some financial backing, incorporated and built a small factory in Madison to produce stationary engines. Their little factory producing stationary engines grew, and in 1900 Hart and Parr added a second building, which they used as a foundry. The partners started working to develop a more powerful self-propelled internal combustion traction engine to move heavy loads. This idea was revolutionary; because no previous internal combustion engines could move their own weight, much less pull something else. Hart and Parr also were known in the tractor industry for pioneering the "two-lunger" or double-lung horizontal engine. These were designed to run on gasoline. According to Hart-Parr records, about 1,200 stationary engines were built at their Madison factory.

As demand for their engines grew, and as they realized the need for larger facilities to build traction engines, Hart and Parr needed capital for expansion again but land was expensive in Madison and the city did not encourage industrial development. They were unable to obtain the needed capital in Madison, but with help from Charles Hart's father, Oliver Hart, they were able to find necessary investors in Hart's hometown of Charles City, Iowa. A group of local businessmen, including the Ellis brothers (attorneys and owners of two local banks), offered a reasonably priced site and additional capital. To obtain that expansion capital, they agreed to move their operations and build their new factory in Charles City. Ground was broken on July 5, 1901, five buildings were constructed and transfer of operations was completed on Dec. 25, 1901.

Charles Hart, the creative leader of the company, was the first president and general manager of the Hart-Parr Co. The advertising manager of the company described Hart as the "dynamic organizer" and Charles Parr as the "practical engineer." The first gasoline traction engine was completed at the new factory in 1902. In 1903, they perfected the first known method of burning kerosene as a fuel, cutting fuel costs in half. Hart-Parr Company continued to build gasoline engines which were not self-propelled, at least until 1903. However, in 1904, the company began to faze them out, and decided to build traction engines exclusively.

By 1907, their company was well-established, and Charles Parr stayed with the company until 1923. From November 1923 to July 1924, Parr was chief engineer for the Elgin Street Sweeper Company. However, he then returned as an engineer for the Hart-Parr Company. He remained with the company when they merged with three other companies in 1929 to become Oliver Farm Equipment Company, and he stayed with that company until he died during a trip to Los Angeles in 1941. Charles Hart had died four years earlier in 1937.

As the builders of the first commercially successful gasoline traction engines or "tractors" powered by internal combustion engines, they have been called the "Founders of the Tractor Industry" by the American Society of Agricultural and Biological Engineers and the American Society of Mechanical Engineers.

Arthur Clifford Howard (1893-1971)

Arthur Clifford Howard (1893-1971) was born on 4th of April 1893 at Crookwell, New South Wales, Australia, eldest son of John Howard, a farmer, and his first wife Mary Ellen, née Smith. After attending schools at Crookwell and Moss Vale, he studied engineering through a correspondence course while an apprentice at Moss Vale.

Encouraged by his father, who had brought the first steam tractor to Gilgandra where he had taken up a property, Howard began in 1912 experiments in rotary tillage which culminated in his invention of the rotary hoe. Using various pieces from farm machinery, he rigged a drive from the tractor engine to the shaft of a one-way notched disc cultivator. He found that the ground could be tilled without the soil-packing that occurred with normal ploughing. But the fast-spinning discs threw the soil sideways until he developed an L-shaped blade mounted on widely spaced flanges fixed to a small-diameter rotor.

Work on a larger model was interrupted by World War I. Unfit for active service; Howard went to England under a scheme initiated by (Sir) Henry Barraclough to work on munitions and aircraft engines. Unable to interest English agricultural implement firms in his ideas, Howard returned to Moss Vale in 1919. Next year he tested and patented his rotary hoe cultivator. It consisted of a main frame carrying an internal combustion engine and a subsidiary frame carrying five rotary hoe cultivators. He continued to develop his ideas, building models to suit particular terrains and types of farming, a rotavator to fit a Fordson tractor and several types of hand-controlled machines. His DH22 tractor, designed in 1927 to work with rotary hoes, initiated the first large-scale production of tractors in Australia.

Howard had married Daisy May Hayes at the Methodist Church, Moss Vale, on 19 September 1925. After moving to England, they lived at Upminster, Essex. A determined, quiet but outgoing man, devoted to his family, Howard was a practical engineer who combined business acumen with vision and outstanding inventive ability. As well as being managing director of Rotary Hoes Ltd until 1970, he was a director of G.D.H. Ltd, Harleston Industries Ltd, Howard (Forge & Foundry) Ltd and Howard Rotavator Co. Ltd. In 1970 he was appointed a Commander of the Order of the British Empire (C.B.E.).

Howard died on 4th January, 1971 at Harold Wood Hospital, Essex, leaving an estate in England valued for probate at £111,476. He was survived by his wife, two daughters and a son, who became managing director of Howard Rotavator Pty Ltd in Australia.

Frank Zybach (1894-1980)

Frank Zybach was born in Oregon in 1894. His family moved to Nebraska later that year. Zybach grew up in Nebraska He began farming in Colorado, but enjoyed inventing and designing much more. In 1947 when he saw a demonstration of modern movable irrigation in which workers were moving and connecting pipes fitted with sprinkler heads from one part of a field to another. But Zybach, a lifelong tinkerer, saw something more: Why have humans set up, take down, move the equipment and repeat? Why not have the equipment move itself?

He invented the self-propelled sprinkling apparatus. Zybach built his first prototype within a year. It rotated around a center wellhead. Guy wires that were attached to support towers held the sprinkler-fitted water pipes above the ground. Control wires and two-way water valves kept the towers in line. The first support towers moved on skids, but Zybach soon replaced those with wheels propelled by the irrigation water itself.

He applied for a patent for the "Zybach Self-Propelled Sprinkling Apparatus" in July 1949. He knew he needed to improve his invention — making it tall enough to work for corn, among other things. So, the same year he got his patent, he moved back to Nebraska and went into business with a friend, A. E. Trowbridge. The duo didn't immediately succeed, partly because Zybach kept making improvements before Trowbridge could sell the models they had already manufactured. They sold the patent rights for a 5 percent royalty to farm-equipment manufacturer Robert Daugherty of Valley Manufacturing (later Valmont) in 1954.

Frank Zybach farmed in Nebraska for the rest of his life. He died in 1980.

8.2 The agrarian revolutionists: their enterprises and growth

Agricultural machinery companies before and during the 1700s

The first known recorded commercial agricultural based enterprise was a flour mill in the timeline of Farmstead, electrification and processing equipment development which was built in the American colonies of New Amsterdam (New York) in 1626. A century later, at about 1783, the earliest known record of the establishment of agricultural machinery manufacturing company was that of a factory for making ploughs in England.

Agricultural machinery companies in the 1800s

By 1827, a hay rake manufacturing business was established by William Stedman, which greatly expanded through the production of hay rakes and other implements by his son Marshall W. Stedman (1859-1935). Between 1830 and 1840, there was a massive shift from blacksmithing to specialized factories dedicated to the manufacturing of reapers, ploughs, threshing machines, and other farm implements. Twin brothers; Hiram A. Pitts (1799-1859), and John A. Pitts (1799-

JOHN A. PITTS

1859), of Maine formed Pitts Agricultural Works in 1837, the company was later named Buffalo Pitts Co., Buffalo, NY in 1877.

John Deere started his business exploits by setting up a blacksmith shop in various towns around Vermont producing ploughs before 1837. Facing a tough business environment, in 1837, the 33-year-old Deere packed up and headed west, eventually settling in Grand Detour, Illinois. There, he set up another blacksmith shop. Deere, beginning with the invention and patenting of steel plough with saw-blade steel and smooth wrought iron; built in 1838.

Increasing demand in 1843 led Deere to partner with Leonard Andrus to produce more ploughs, and by 1846, production had risen dramatically—that year, Deere and Andrus produced nearly 1,000 ploughs. The following year, Deere decided that Grand Detour, Illinois, was lacking as a hub of commerce, so he sold his interest in the blacksmith shop to Andrus and moved to Moline, Illinois, located on the Mississippi River. There, he was able to offer the advantages of water power and cheaper transportation. Deere soon began importing British steel, which successfully speed up manufacturing—his company made 1,600 ploughs in 1850, and began producing other tools to complement its line of ploughs. Deere's next move was to contract with Pittsburgh manufacturers to develop comparable steel plates, thereby avoiding the troubles of overseas importation. By 1857, John Deere's factory was selling over 10,000 steel ploughs a year.

In 1868, John and his partners found and incorporated, Deere & Company, which is still in existence today. Deere & Company in Moline, Illinois, manufacturer of a full line of John Deere implements, had been watching the progress of the Waterloo Engine Company and the increasing quality of its products. In 1918, the John Deere plough-manufacturing company bought Waterloo for $2,350,000. The company became the largest combine manufacturer in North America, and expanded into a full line of tillage and planting equipment manufacturing. By 2012, the company's worth had climbed to more than $40 billion.

By 1831, the invention of mechanical reaper with serrated cutting bar that moved back and forth, invented by the son of Robert McCormick (1780-1846), Cyrus H. McCormick Sr. (1809-1884) of Walnut Grove, VA, which was patented in 1834 led to the establishment of McCormick factory. Cyrus followed the work of his father and the business, McCormick Reaper Works, moved from Virginia to Chicago to establish its first Chicago factory to produce reaper machines in 1847 and by 1848, Nichols & Shepard Company

(USA) initiated the production of a threshing machine that evolved into the popular Red River Special. His company became a part of Oliver Farm Equipment Co. later in 1929. The firm was incorporated in later years as the Empire Drill Co.

The origin of McCormick Reaper Works rested on the inventive abilities of Cyrus McCormick, Sr., in combining the seven elements basic to any successful reaper, as a company it would rarely be a leader in marketing new products. With its established brand name, and solid marketing organization, it had little reason to push new designs or new machines; when others made important innovations, however, McCormick was almost always well prepared with patents and expensive lawyers to help it produce a competitive product.

After the death of Cyrus, Sr., in 1884, Cyrus, Jr., renovated the marketing organization and focused manufacturing capabilities on producing a durable machine with the least complexity, largely limiting the company to a single model of each principal machine -- reaper, self-rake, mower, and binder - to give maximum production runs. By the 1890s McCormick not only commanded a premium price - often 10 percent or more above any rival - but built sturdy machines at lower costs than any competitor. One of the McCormick's greatest competitors was William Deering.

William Deering's contributions to American agricultural equipment development before and after the turn of the 20th century remain an enigma to some collectors. His appearance on the scene was his partnership with Elijah Gammon, providing $40,000 in funding for the production of a horse-drawn grain harvester developed by two brothers; William Marsh and Charles Marsh. By 1872 the company showed $80,000 in profits, and in 1873 the name was changed to Gammon & Deering Co. to reflect Deering's management role. By 1879 Deering was the sole owner and the company's name had been changed to Deering Manufacturing Company.

The illness of his partner, Gammon left him in charge of the business. By 1879 Deering was the sole owner and the company's name had been changed to Deering Manufacturing Company. He soon purchased the Appleby twine-knotter patents and along with the Marsh harvester, the company pioneered a harvesting reaper incorporating an automatic twine binder based on John Appleby's invention. Deering Manufacturing Company produced and sold 3,000 of Appleby's twine-tie binder for the 1880 harvest, with profits above $400,000. In 1880, Deering moved the company to Chicago and established the Deering Harvester Works.

McCormick was never far behind Deering in making improvements and often profited from Deering's experiences. It was the McCormick vertical corn harvester that became the industry standard, as did the McCormick design for roller bearings. McCormick built its own twine mill in 1900, more than 15 years after Deering, but it had in fact owned a major

Boston mill in the 1880s and had secretly occupied a strong position in the sisal trade for a decade; Deering had, unknowingly, bought much of its raw material for twine from McCormick.

By this means and others the McCormick's always seemed to have had full knowledge of their competitors' activities; they never feared these competitors, but they did find Deering "troublesome." Undoubtedly the Deering move into coal, ore, and steel aggravated that feeling; such expansion demanded heavy expenditures. McCormick was investing heavily to expand its Chicago factory to make most machine components and to develop its foreign trade, where it probably had a much larger share of the trade than in America. It would have exposed the McCormicks to heavy financial pressure to follow the Deerings simultaneously into full vertical integration.

However, the McCormicks were under less pressure than the Deerings; they had only just begun producing most of their own components, but had a long, stable relationship with Jones and Laughlin. Even so, the new purchasing department had in 1900 begun dividing most steel orders between competing suppliers, a strategy which avoided the problems Deering had faced. Gary's suggestion nevertheless struck a responsive chord with the McCormicks, perhaps a responsiveness enhanced by Gary's long association with the Deerings for whom he had been principal legal counselor before leaving legal practice in 1898 to head Federal Steel. Cyrus H. McCormick, Jr. led the McCormick Harvesting Machine Co. formed in 1902. By 1902, McCormick Co. probably commanded 40 percent of the harvester trade.

At the time of McCormick's exploits, at another part of America were two brothers working on developing a method of binding freshly reaped wheat stalks; Charles Marsh and William Marsh of Illinois, converted the McCormick reaper into the world's first harvester. Marsh Brothers was founded in 1858. In 1859, the Marsh brothers unveiled the completed product during a reaping contest north of DeKalb. William Marsh went to Piano, Ill., in the winter of 1860, where he formed a company; Steward & Company, with a prominent settler, Lewis Steward, who invested in the company to build the Marsh harvester.

The Marsh brothers withdrew from the Plano group at about 1869, and started their own plant in Sycamore, Ill. Elijah H. Gammon, a relatively wealthy man and former Methodist preacher, were soon involved with the firm in the 1870s. In 1870, their business got an added financial boost and a new partner in the person of William Deering two years later. William Deering purchased the company in October 1875 and moved most of its operations to North Chicago. It merged with the McCormick Harvesting Machine Company in 1902 to form International Harvester.

In 1854, H. L, & C. P. Browns established a factory under the firm name of H. L, & C. P. Brown Shortsville to manufacture seed drills. In 1867 C. E. Patric who had been in the employ of the Browns, moved to Springfield Macedon, N. Y., where he and Lyman Bickford took out several patents in 1867, covering the "double distributor planter." The distinguishing feature of this invention was a seed-wheel or disk with carrying flanges on each side, one chamber feeding coarse, bulky seeds, like oats, and the other being smaller, to sow wheat, rye, etc. The invention was adopted by Bickford & Huffman, of Macedon, and licensed Ferrell, Ludlow & Rodgers, later Thomas, Ludlow & Rodgers, incorporated in 1883 as the Superior Drill Co.

Alanson Harris (1816-1894), and his son John Harris established an implement manufacturing and sales company in Beamsville, Ontario, Canada in 1857. The company was renamed the Massey Manufacturing Co. in 1878, and later moved to Toronto, Canada. In 1852, M. & J. Rumely firm started by Meinrad Rumely (1823-1904), who purchased John Rumely's share in 1882, incorporating as M. Rumely Co. in 1887, in La Porte, IN. In **1863** Early steam engine built by M. & J. Rumely Co., La Porte, IN

Johnston Harvester Co. was established by Samuel Johnston (1835-1911), Brockport, NY in 1871. It made reaping machines, including self-raking devices (1856ff.), and patented rotary and disc harrows. This company succeeded Johnston, Huntley & Co., Syracuse, NY. Joseph F. Glidden (1813-1906), with Isaac Ellwood (1833-1910), formed Barb Fence Co. in 1874, in Illinois to produce barbed wire after their patient in 1844.

In 1865, Aultman and Taylor Machinery Co. founded in Mansfield, OH built steam engines and threshers; Advance-Rumely Co. obtained company assets in 1923. One of the first commercially successful steam traction engines was built by C. & G. Cooper Co. in **1876**

DANIEL MASSEY

In 1877, Empire Drill Co. was formed to manufacture planting equipment; it was however purchased by the American Seeding Machine Co. in 1903. Brantford Plough Works was founded in Brantford, Canada, in 1877 to build tillage implements by family of James G. Cockshutt, (1853-1885). Brantford Plough Works transformed to a Company named Cockshutt Plough Co. Ltd. In 1882 and expanded into tractor production, building the first tractor with live PTO in 1924. Ohio Cultivator Co. was founded by Harlow Case Stahl (1850-1941) in 1878, established in Fremont, OH, and later moved to Bellevue, OH.

In 1880 Marsh reapers built at Marsh factory in Plano, IL, which later became the firm of Gammon & Deering, controlled by Elijah H. Gammon (1819-1891), and William

Deering,1826-1913. In 1882 Meinard Rumely (1823-1904), bought interests of M & J Rumely (John Rumely, 1853-1931) and established M. Rumely Co. William Deering & Co. was formed to manufacture farm implements in 1883, after being known as Deering Harvester Co., organized in 1880. Robert H. and Cyrus M. Avery, Galesburg, IL formed Avery Co. Peoria, IL (originally called Avery Planter Co., founded in 1874) in 1891 to build steam traction engines. Same year 1891, Hornsby-Ackroid oil engine was patented and originally manufactured by Richard Hornsby & Sons of Grantham, England

Massey-Harris Co., Ltd., was formed in 1891 in Canada, by merging the Massey Manufacturing Co., Toronto, and Alanson Harris, Son & Co., Ltd., Brantford, Canada, with headquarters in Toronto, Canada (The company existed until 1953, became Massey- Harris-Ferguson Co., Ltd., and in 1958, became Massey-Ferguson). Several company acquisitions followed the merger and by 1891 Sidehill combine was introduced by Stockton Wheel Co., CA, by Holt Brothers, led by Benjamin Holt (1849-1920).

In 1893 Waterloo Gasoline Traction Engine Co. was incorporated in Waterloo, IA, by John Froelich; company reorganized as Waterloo Gasoline Engine Co., Waterloo, IA, in 1895 as one of several leaders in developing and building gasoline agricultural traction engines. The first prototype built for John Froelich in 1892 had a 20 hp engine built by Van Duze Gas and Gasoline Engine Co., OH. The Froelich traction engine set the stage for the Waterloo Boy and the John Deere line of traction engines. In 1897 Hart-Parr Co., Madison, WI, reorganized and moved to Charles City, IA in to build traction engines. Huber Manufacturing Co., Marion, OH, which had been manufacturing steam traction engines, also entered the manufacture of gasoline traction engines in 1898.

Agricultural machinery companies in the 1900s

In 1901 Allis-Chalmers Co. was formed by the merger of Edward P. Allis Co. Milwaukee, WI (a company named by Edward P. Allis, Jr.(1849-1918)following death of his father, Edward P. Allis in 1889), and Chalmers & Co. (formed by the father of William J. Chalmers (1852-1938) and other companies. David Pryce Davies (1870-1948), the designer of Case side-crank steam engine joined the Edward P. Allis Co. in 1898. He joined the Allis-Chalmers Co. in 1903, and in 1910 returned to the J. I. Case Co. to design traction engines and develop gasoline engines.

After several years of successful innovations using steam, Charles W. Hart (1872-1937), and Charles H. Parr (1868-1941), built their first factory, Hart-Parr Co. in USA in 1902 to manufacture gasoline traction engines driven by an internal combustion engine, adapted the internal combustion Otto cycle engine that could propel farm equipment. Same year, 1902 recorded the birth of International Harvester Co. Eight companies made up the merger group incorporating William Deering Harvester & Co., Chicago, IL; McCormick Harvesting Machine Co., Chicago, IL; Warder, Bushnell, Glessner & Co.; Plano Manufacturing Co., Plano, IL; Milwaukee Harvester Co., Milwaukee, WI; Osborne line of farm machinery and Champion Reaper Works.

The IHC market share of the total farm machinery was 85 percent. After forming IHC, several new lines acquired including Minneapolis Harvester Co., whose first harvester was built in 1883, in 1903; Weber Wagon Co., IL, acquired in 1903; David M. Osborne & Co., NY, and Aultman-Miller Co., Canton, OH, acquired in 1904; the Keystone Co., Rock Falls, IL, hay tools and corn shredder (originally introduced in 1872); and Kemp manure spreader. McCormick was the oldest and largest of the harvester manufacturers, having begun regular production in 1842.

Hart and Parr (Charles Walter Hart and Charles Henry Parr) incorporated their first business, Hart-Parr Co. in 1905, specializing in manufacture of traction engines; used the word "tractor" beginning in 1906 instead of "gasoline traction engine."

By 1907, their company was well-established, and a Hart-Parr sales manager combined the words traction and power to coin the term "tractor" and replace the clumsy term "gasoline traction engine" in advertisements. By 1915, the Hart-Parr Company was capitalized at $2.5 million, and was one of America's major farm equipment manufacturers. After Charles Hart left the Hart-Parr Company in 1917 for unascertained reasons, Charles Parr stayed with the company until 1923. From November 1923 to July 1924, Parr was chief engineer for the Elgin Street Sweeper Company. However, he then returned as an engineer for the Hart-Parr Company. He remained with the company when they merged with three other companies in 1929 to become Oliver Farm Equipment Company.

In 1909, Avery Co. (originally named the Avery Farm Machinery Co., and then the Avery Power Machinery Co.), Peoria, IL, manufactured a tractor truck,12-36 hp. In 1922 Avery Co. made a caterpillar-type tractor called the Track-Runner. The company originally made cultivating tractors, steam engines, and threshers.

In 1910, C. L. Best Gas Traction Co. (later called C. L. Best Traction Co.) was formed in Elmhurst, CA, by Clarence L. Best (1875-1951), the son of Daniel Best (1838-1923); The Company introduced the track-type tractor in 1911. In 1925 this company merged with the Holt Manufacturing Co. to form the Caterpillar Tractor Co.

M. Rumely Co. began to manufacture threshing machines in 1911. M. Rumely Co. acquired the Advance Thresher Co., established in 1885, and the Gaar-Scott & Co., Richmond, IN, established in 1836, and formed the Advance Rumely Co. in 1915. Allis-Chalmers Co. acquired most of the assets of the Advance Rumely Co. in 1931.

C. L. Best Gas Traction Co. (formed by Clarence L. Best, 1875-1951, son of Daniel Best) with name later changed to C. L. Best Tractor Co. In 1912, Best Co. began production with 70 hp track-laying tractor while Hornsby and Sons of Great Britain, sold their interest in caterpillar-type tractors to Holt Manufacturing Co., Stockton, CA. In 1913, Allis-Chalmers Manufacturing Co. was incorporated in 1913, building on the Allis-Chalmers Co. formed in 1901, acquired several companies over the years, including the Advance-Rumely Co. in 1931.

In Australia, Arthur Clifford Howard (1893-1971), with a fellow apprentice, Everard McCleary (who died in 1918), established a company to manufacture rotary hoe cultivators in 1912, but their plans were interrupted by World War I. In 1919 Howard returned to Australia and resumed his design work, patenting a design with 5 rotary hoe cultivator blades and an internal combustion engine in 1920 (Howard, 1983). In March 1922 Howard formed a syndicate, Austral Auto Cultivators Pty Ltd, which later became known as Howard Auto Cultivators based in Northmead, a suburb of Sydney from 1927 Ltd (Howard, 1983).

The firm moved to Northmead, Sydney, in 1927. Despite reduced sales during the Depression, Howard raised new capital for the company, now known as Howard Auto Cultivators. Finding it difficult to meet overseas demand stimulated by a successful export drive, he arranged a ten-year license with an English firm to make his machines for markets outside Australia. In 1937 unauthorized design alterations took Howard to England where the license was terminated. Howard was finding it increasingly difficult to meet a growing worldwide demand for exports of his machines. He travelled to the United Kingdom, founding the company Rotary Hoes Ltd in East Horndon, Essex, in July 1938 (The Howard Rotavator, 2007).

He returned briefly to Australia to resign his position of managing director at Howard Auto Cultivators and in July 1938 formed an English company, Rotary Hoes Ltd, which established branches in the United States of America, South Africa, Germany, France, Italy, Spain, Australia and New Zealand. It later became a holding company for a wholly owned subsidiary, Howard Rotavator Co. Ltd, to which it transferred the manufacture and distribution of rotavators, manure spreaders, trench diggers and soil stabilization machinery. The company received the Queen's award to industry in 1966. The Howard Group of companies was acquired by the Danish Thrige Agro Group in 1985, and in December 2000 the Howard Group became a member of Kongskilde Industries of Soroe, Denmark (Machinery Manufacturers, 2006).

Drott Tractor Co. was established by Edward A. Drott (1887-1956) in 1916 to expand the use of the crawler tractor, particularly for logging operations. He developed the skid-loader. 1916 witnessed the marketing of crawler tractor, Model R by Cleveland Motor Plough Co., Cleveland, OH. The company was renamed Cleveland Tractor Co. in 1917, of which Rollin H. White (1872-1962), was president until 1929 (name Cletrac trademarked in 1919). After 1923, Cletrac tractor developed by Cleveland Tractor Co. by William King White, 1901-1947. His father, Rollin H. White, 1872-1962, invented the White Steamer (automobile and truck) in 1899.

In 1918 Osborne Co. line of equipment, primarily leaders in the development of mowers, reapers, and binders, originated by David M. Osborne, 1822-1866, was sold to International Harvester Co. Henneuse Tractor Co. was formed in 1920 by Clarence A. Henneuse, 1879-1939, to design crawler tractors in Bucyrus, OH, , later sold to Huber Manufacturing Co. of Marion, OH. Caterpillar Tractor Co. was formed in 1925, while incorporating Holt Manufacturing Co. and Daniel Best (Daniel Best, 1838-1923, followed by Clarence L. Best, 1875-1951, of C. L Best Gas Traction Co.)

Wallis tractor Co. of 1912 was absorbed by Massey-Harris Co. in 1928. In 1929, Oliver Farm Equipment Co. (Oliver Chilled Plough Co. founded in 1853), headed by Cal Sivright, 1886-1945 merged and absorbed Hart-Parr Tractor Works Co., Nichols & Shepard Threshing Machine Co. (formed in 1886), and American Seeding Machinery Co. (formed in 1850).

Minneapolis-Moline Power Implement Co. was formed in 1929 with the merger of three companies: Moline Plough Co. (1870-1929), Minneapolis Steel & Machinery Co. (1902-1929), and Minneapolis Threshing Machine Co. (1874-1929). In 1935, Caterpillar Co. discontinued the combine business and sold combine line to Deere and Co.

In 1929, Minneapolis-Moline Power Implement Co. was formed by union of Minneapolis Steel & Machinery Co. (founded in 1902)., Minneapolis Threshing Machine Co. (founded in 1887), and Moline Plough Co. (founded in 1870), by James L. Record, 1857-1944. Name changed to Minneapolis-Moline Co. in 1949 (see additional information in the Implements timeline). H. D. Hume Co. was established in Mendota, IL (formerly a part of Hume-Love Co.) in 1941 by Horace D. Hume and James Edward Love who formed a partnership (Hume-Love Co., Garfield, WA)

Cleveland Tractor Co., builder of Cletrac, merged with Oliver Farm Equipment Co., became the Oliver Corporation in 1944. W. (William) King White, 1901-1947, guided development of Cletrac tractor. The Oliver Farm Equipment Co. was absorbed by the White Motor Co. in 1945 (Cal Sivright, 1886-1946, W. King White, 1901-1947, Rollin H. White, 1872-1962, and his brother Clarence White). Ann Arbor Manufacturing Co. originally in Ann Arbor, MI, commemorated at Shelbyville, IL became part of Oliver Farm Equipment Co. in 1953, and later the White Motor Corp.

National Farm Machinery Cooperative (NFMC) formed in 1945, including at least ten USA states with Farm Bureaus and a similar organization with the Canadian Cooperative Implements Limited (CCIL), farmer-owned organizations for distribution of tractors and implements. Gleaner Harvester Works responsible for development and patent of the Baldwin Gleaner combine noted as the first combine in North America to offer a corn head was acquired by Allis-Chalmers Co. in 1955.

American Tractor Corp. was formed by the Rojtman family, headed by Marc B. Rojtman (1917-1967), manufacturing the GT-25 crawler in 1949 and with the TerraTrac trademark. Terratrac Gasoline Crawler Tractors (GT) and Diesel (DT) built by American Tractor Corp., Churubusco, IN, later (1957) merged with Case Co. and moved to Burlington, IA. In 1951, American Tractor Corp. produced the GT-30 (NTT 471). The company merged with J. I. Case in 1956. Same year, Deere & Co. opened Research & Engineering Center devoted exclusively to design and testing tractors, in Waterloo, IA.

Massey-Ferguson Co. was formed in 1958 by the merger of Harry Ferguson firm with Massey-Harris Co., following a brief period as the Massey-Harris-Ferguson organization and in 1959; Massey-Ferguson Co. acquired F. Perkins, Ltd. originally formed in 1932, Peterborough, England, which produced diesel engines. In 1978 all Massey-Ferguson combines were diesel powered. .

In 1960, White Motor Co., a builder of trucks in Cleveland, OH, USA acquired Oliver Corp., which was consolidated at Charles City, IA, for tractors; in South Bend, IN, for ploughs and tillage tools; in Shelbyville, IL, for haying equipment. White Motor Corp. acquired Cockshutt Farm Machinery Co., Ltd., Brantford, Canada, and combined with Oliver Corp. in 1962. Minneapolis-Moline Co., Hopkins, MN, became a subsidiary of

White Motor Co. in 1963 and in 1964; J. I. Case Co. acquired the Colt Manufacturing Co., Winneconne, WI, for production of small lawn and garden tractors, 10-12 hp.

By 1969, Minneapolis-Moline Co. was joined with Oliver Farm Equipment Co., (a subsidiary of the White Motor Co.) and the Cockshutt Equipment Co. Ltd. of Canada to form the White Farm Equipment Co., Oak Brook, IL. In 1970, J. I. Case Co. became a wholly-owned subsidiary of Tenneco, Inc., Houston, TX. By 1985, J. I. Case Co. and International Harvester tractor lines combined under the control of J. I. Case Co.; tractors called Case International or Case-IH. Also in 1985, Ford Motor Co. purchased the New Holland division of Sperry Corp. to form a full line farm tractor and equipment company, with tractors known as Ford New Holland. Same year, Allis-Chalmers Co. was sold to Klockner-Humboldt-Deutz (KHD), AG, Cologne, West Germany.

In 1990 Allis-Gleaner Co. formed AGCO, originally including Deutz-Allis Corp. of North America and Klockner-Humboldt-Deutz (KHD) of Germany, including tractors and equipment, later acquiring Hesston, McConnell, White, Massey Ferguson, Tye, Flencoe, Farmhand, and other companies. AGCO purchased the White-New Idea line of products from Allied Products and North American Ferguson distribution in 1993 and the Massey Ferguson division of Varity Corporation (not including the Perkins Engine Group) and the McConnell 4WD tractors in 1994.

References

Agriculture in the Bible Encyclopedia - ISBE (Bible History Online)_files\bhologo.jpg

Ammann K. 2007. Reconciling Traditional Knowledge with Modern Agriculture: A Guide for Building Bridges. In *Intellectual Property Management in Health and Agricultural Innovation: A Handbook of Best Practices* (eds. A Krattiger, RT Mahoney, L Nelsen, et al.). MIHR: Oxford, U.K., and PIPRA: Davis, U.S.A. Available online at www.ipHandbook.org.

Anushri, main features of an agricultural society. URL:www.preservearticles.com/201107058901/4-main-features-of-an-agricultural-society.html

Amanda Etty, Sow your seeds with a dibber. *http://www.canadiangardening.com/*

Ardrey, Robert L. 1894 *American Agricultural Implements: A Review of Invention and Development in the Agricultural Implement Industry of the United States* . Chicago: R. L. Ardrey, 1894 *http://dig.lib.niu.edu/*

ASME, 1996. The Hart-Parr Tractor Introduced In 1901. A National Historic Mechanical Engineering Landmark. The American Society Of Mechanical Engineers Charles City, Iowa. May 18, 1996

ASME, 2014. Rumely Companies' Agricultural Products 1908. https://www.asme.org/

Astrand, B., and A. Baerveldt. 2002. An agricultural mobile robot with vision-based perception for mechanical weed control. *Autonomous Robots* 13: 21-35.

Austin Gingerich, 2009. History of Deere.

Baker D. G. A brief excursion into three agricultural revolutions. Kuehnast lecture - number IV

Baker, D.G., D.L. Ruschy, and R.H. Skaggs. 1993. Agriculture and the Recent "Benign Climate" in Minnesota. Bull. Amer. Meteor. Soc. 74:1035-1040Burke, J. 1978. Connections. Little, Brown and Co., Boston. Pp. 304Gimpel, J. 1976. The Medieval Machine. The Industrial Revolution of the Middle Ages. Holt, Rinehart and Winston, N.Y. Pp. 274.

Bakker, T. 2009. An autonomous robot for weed control - design, navigation and control. PhD diss. Wageningen, The Netherlands. Wageningen University. Department of Agricultural Engineering.

Balter, Michael. 2007. Seeking Agriculture's Ancient Roots. *Science* 316 (5833): 1830–1835.

Barlow, Robert Stockes; "300 Years of Farm Implements and Machinery 1630–1930"; Krause Publications (2003); p.33; ISBN 978-0873496322

Bello, R. S. 2007. Fundamental Principles of Agricultural Engineering Practice. Climax Printers, Enugu Nigeria

Big Era Three Farming and the Emergence of Complex Societies 10,000 - 1000 BCE http://worldhistoryforusall.sdsu.edu/default.php

Bill Ganzel, 2007. Farming & Rural Life in the 1950s & 60s. http://www.livinghistoryfarm.org /farminginthe50s /farminginthe1950s.html

Bill Ganzel, 2007. Postwar Technology. http://www.livinghistoryfarm.org/farminginthe40s /farminginthe1940s.html

Bill Ganzel, 2007. Farming & Rural Life in the 1970s to Today. farminginthe1970s_files\farmingin the 70s-title.gif

Bill Lenox , 1978. FARM Tools through the ages.http://www.farmcollector.com/ March/ April1978

Bill Vossler, 2008. Froelich launched New Era on the farm. Copyright 2014, Ogden Publications, Inc., 1503 SW 42nd St., Topeka, Kansas 66609-1265. http://www.farmcollector.com/

Bob Barton, 2010. Richard Trevithick – 200 Years of Steam Trains. URL: http://www.ego4u.com/en/read-on/countries/uk/richard-trevithick. Changed: 10th Dec 2010 19:39

Bond, W., R.J. Turner and A.C. Grundy.2003.A Review Of Non-Chemical Management. Coventry, UK. Aval. at www.gardenorganic.org.uk/organicweeds /downloads/updated_review.pdf. Accessed 6 September 2010.

Bowman, G. 1997. Steel In The Field: A Farmer's Guide To Weed Management Tools. Beltsville, Maryland. Sustainable Agriculture Network Handbook Series; 2.

Braidwood, Robert J. 1960. The Agricultural Revolution. *Scientific American* 203(3): 130–148.

Britta K. Ager, 2012. Roman Agricultural Magic (Article)- Ancient History Encyclopedia. htm Published on 19 April 2012

Brown, Tony (1997). "Clearances and Clearings: Deforestation in Mesolithic/Neolithic Britain". *Oxford Journal of Archaeology* 16 (2): 133. doi:10.1111/1468-0092.00030.

Carl W. Hall. 2011. Part 2. Timeline in the development of agricultural field implements, related apparatus, and equipment. Timelines in the Development of Agricultural and Biological Engineering. © ASABE, St. Joseph, MI.

Carl W. Hall. 2011. Part 1. Timeline in the development of agricultural tractors and power units. Timelines in the Development of Agricultural and Biological Engineering. © ASABE, St. Joseph, MI

Carl W. Hall. 2011. Part 3. Timeline in the development of farmstead, electrification, and processing equipment . Timelines in the Development of Agricultural and Biological Engineering. © ASABE, St. Joseph, MI

Christian Williams, 2012. Van Duzen: Whatta Doozy! December 2012 2011/January. http://www.gasenginemagazine.com/

Cloutier, D C, R. Y. V. D. Weide, A. Peruzzi, and M. L. Leblanc. 2007. Mechanical weed management. In *Non-Chemical Weed Management.* Upadhyaya M.K and R E Blackshaw, ed.111-134. CAB International

Colin Schultz 2013. Early Agriculture Nearly Tanked Ancient Europe's Population. http://www.smithsonianmag.com/smart-news/

Colin Stief, 2014. Slash and Burn Agriculture. About.com. \od\geographyintern\a\colinbio.htm. ©2014

Constable, George; Somerville, Bob (2003). *A Century of Innovation: Twenty Engineering Achievements That Transformed Our Lives, Chapter 7, Agricultural Mechanization.* Washington, DC: Joseph Henry Press. ISBN 0-309-08908-5.

David Schnakenberg, 2013. The Marsh Harvester. http://www.farmcollector.com/ Farm Collector March 2013

David W. Koeller, 1998. www.thenagain.info/webchron/glossary/earlyag.n.html

Dairy Waste Management Systems http://www.demographia.com/dm-lon31.htm population list on demographia.com, Accessed 15 Jan 2014

De Datta SK, Tauro AC, Balaoing SN., 1968. "Effect of plant type and nitrogen level on growth characteristics and grain yield of indica rice in the tropics". *Agron. J.* 60 (6): 643–7.

Donald G. Baker, 2014. Kuehnast lecture - Number IV: A brief excursion into three agricultural revolutions. http://climate.umn.edu/doc/journal/kuehnast_lecture/kuehnast.htm

Donald Johnson, 2014. Agricultural Mechanization Essay. http://www.greatachievements.org/?id=2955

Dorrington P. C., 2004. Antique farm tools (used mainly in England, Wales and Scotland). http://www.mrsite.com/

Economics-papers:The-basic-characteristics-of-modern-agriculture-and-the-focal-point eng.hi138.com/economics-papers/industry-economic-papers/200512/45571_the-basic-characteristics-of-modern-agriculture-and-the-focal-point.asp#.UtYEXKNxjIU

Ecifm, Agriculture in post war Britain.. http://www.ecifm.rdg.ac.uk/directory.htm

Editors of Encyclopædia Britannica, 2014. Obed Hussey, http://www.britannica.com/

Edward P. Allis and Company Reliance Works. Illustrated Catalogue of Roller Mills and other Special Machinery." (Milwaukee: Cramer, Aikens, Engravers and Printers, 1888); Online facsimile at: http://www.wisconsinhistory.org/turningpoints/search.asp?id=1186; Visited on: 2/18/2014

Encyclopedia Britannica Online. "hand tool: Neolithic tools". Retrieved on 2014-01-27.

Fajemirokun. F. A. and Olusegun T. Badejo, 2004. A New Curriculum for Surveying Education in Nigeria, 3rd FIG Regional Conference Jakarta, Indonesia, October 3-7, 2004

Fajemirokun. F. A., 1983. The Profession of Surveying in Nigeria and its Contributions to National Development. Second Annual Olumide Memorial Lecture, Nigerian Institution of Surveyors, University of Lagos. October 17.

FAO, 1997. Agricultural Operations Technology For Small Farmers In Eastern And Southern Africa. Terminal Report, FAO/Government Cooperative Programme GCP/RAF/314/SW, Rome 1997

FAO, 1997. Contribution of farm power to smallholder livelihoods in sub-Saharan Africa. Produced by: Natural Resources Management and Environment Department. http://www.fao.org/documents/

Farm Collectors, 2014. Why Did Charles Hart Leave Hart-Parr? © 2014, Ogden Publications, Inc., 1503 SW 42nd St., Topeka, Kansas 66609-1265. http://www.farmcollector.com/

Feng-yun Zhang, 2011. The Basic Path of Modern Agriculture with Chinese Characteristics-Industry Nurturing Agriculture. *Asian Agricultural Research*, 2011, vol. 03, issue 07

Ford, Henry; Samuel (1922). *My Life and Work: An autobiography of Henry Ford*

Flannery, T (1994). *The future eaters*. Melbourne: Reed Books. ISBN 0-7301-0422-2.

Fred Hultstrand and F.A. Pazandak Implements Used on the Farm Mold Plough (Northern Great Plains, 1880-1920). http://memory.loc.gov/ammem/award97/ndfahtml/ngp_farm.html

Gafar Lawal,2013. Farm Mechanization in Nigeria. Posted by Kofo Durosinmi-EttiJuly 31, 2013. http://www2.ca.uky.edu/agripedia/GLOSSARY/farmmech.htm

Gas Engine Magazine, 2012. Circa 1914 Van Duzen http://www.gasenginemagazine.com/. August /September 2012

Gaud William S., 1968. "The Green Revolution: Accomplishments and Apprehensions". AgBioWorld. 8 March 1968 Retrieved 8 Feb 2014.

Georgetown University, 2002. Nomadic vs. Settled Societies in World History. World History Georgetown University, December 2002© JULIA ALLISON 2014http://juliaallison.com/

Gianessi, L. P and Sankula S., 2003. The value of herbicides in U.S. crop production. Washington D.C. National Center of Food and Agricultural Policy.

Greg Wadley and Angus Martin, 1993. The origins of agriculture: a biological perspective and a new hypothesis. Published in Australian Biologist *6: 96-105, June 1993*

Grigg, D.B, (1974). Agricultural Systems of the World. Cambridge University Press. pg. 53-54

Halleym Ned, (1996). Farm. Dorling Kindersky. Old Farm Tools and Machinery, David and Charles Publishers Limited. pg 116

Harrison, L F C (1989). *The Common People, a History from the Norman Conquest to the Present.* Glasgow: Fontana. ISBN 978-0-00-686163-8.

Hart-Parr tractors. © Pioneer Productions. http://www.easycounter.com/

Hazell, Peter B.R. (2009). "The Asian Green Revolution". *IFPRI Discussion Paper* (Intl Food Policy Res Inst). GGKEY:HS2UT4LADZD.

Henry K. Landis, 2013. Never-ending Toil: A Short History of Animal and Human-powered Treadmills September 30, 2013. http://www.blogger.com/profile/17031041618915866221

Herbert Bergmann and Richard Butter, 1985. Primary School Agriculture Volume II: Background Information http://collections.infocollections.org/ukedu/en/(GTZ; 1985; 190 pages)

Hillsdale County Fair, 1851. The Agricultural Society & The Fair. 150 Years in the Hills and Dales, Vol. I. p. 86.http://www.hillsdalecounty.info/history0002.asp

Hillside Cultivator Company.2011. Eco weeder. Available at http://www.hillsidecultivator.com/?page_id=39. Accessed 8 July 2011.

History of agriculture. 2011. http://www.ecifm.rdg.ac.uk/history.htm

History Source LLC 2012. Africa Farming Development. http://historylink101.com/index.htm

Honda Worldwide, 1998| News Release | April 27, 1998 http://world.honda.com/news/1998/p980427a.html

Howard, Arthur Clifford (1893-1971). Australian Dictionary of Biography – Online Edition. Australian National University. 1983. Retrieved 2007-07-12.

Howard F. Stein, 2007. Farming Culture. © Oklahoma Historical Society. http://digital.library.okstate.edu/encyclopedia/index.html

http://www.kuhn.co.uk/uk/range/soil-preparation/folding-power-harrows/hr-6004-r.html

http://www2.ca.uky.edu/agripedia/GLOSSARY/farmmech.htm

http://blogs.worldbank.org/publicsphere/when-robots-attack?cid=EXT_TWBN_D_EXT

http://en.wikipedia.org/wiki/Mechanised_agriculture

http://sunnewsonline.com/new/business/nigeria-tops-countries-with-least-mechanized-farming-hargrave/

http://www.fmard.gov.ng/index.php/issues-in-agriculture/94-the-role-of-mechanization-services-in-the-ata

IITA, 2007. Evolution of tractors. http://old.iita.org/cms/details/index.aspx

Jan van der Crabben, 2011. Agriculture in the Fertile Crescent Article. Ancient History Encyclopedia published on 23 February 2011

James Bowery, 2007. Happy Jethro Tull Day!. Posted on Friday, March 16, 2007. http://majorityrights.com/

Jan van der Crabben, 2011. Agriculture in the Fertile Crescent, 2011. Agriculture in the Fertile Crescent (Article) -- Ancient History Encyclopedia.htm published on 23 February 2011

Jennifer Carrel, 2014. Agriculture Implements From Conner Prairie's Collection. http://www.connerprairie.org/

Joe Morris and Paul J. Burgess, 2012 Environmental Impacts of Modern Agriculture ISBN: 978-1-84973-385-4 eISBN: 978-1-84973-497-4 DOI:10.1039/9781849734974-00001. http://pubs.rsc.org/en/content/ebook/978-1-84973-385-4

Johnson, Paul, C. (1976). Farm Invention in Making of America. Wallace-Homestead Books

Jonathan Bell, "Wooden s From The Mountains Of Mourne, Ireland," *Tools & Tillage* (1980) 4#1 pp 46-56; Mervyn Watson, "Common Irish Types And Tillage Techniques," *Tools & Tillage* (1985) 5#2 pp 85-98.

Ian M. Johnston, 2004 The first tractor engines. DECEMBER 2003-JANUARY 2004. THE AUSTRALIAN COTTONGROWER

Kagan, Donald (2004). *The Western Heritage*. London: Prentice Hall. pp. 535–539. ISBN 0-13-182

Katherine Greening (March 1927). "Early Life of John Francis Appleby". *The Wisconsin Magazine of History* (Wisconsin Historical Society) 10 (3): 310–312. JSTOR 4630672

Kathy Gerkins, 2008. Charles Henry Parr. http://www.findagrave.com/cgi-bin/fg.cgi?page=mr&MRid=39861343 Record added: Dec 28, 2008

Kris Hirst K., 2014. History of Agriculture, Invention of Farming Crops and Raising Animals. http://archaeology.about.com/

Larry O'Dell, 2007. Agricultural Mechanization. © Oklahoma Historical Society http://digital.library.okstate.edu/encyclopedia/toc.html

Lee, Norman, E. (1960). Harvests and Harvesting through the Ages. Cambridge University Press.

Lee, R. B. & DeVore, I., 1968, Problems in the study of hunters and gatherers, in Lee, R.B. & DeVore, I., eds, Man the hunter, Aldine, Chicago.

LeStrange M., Tulare Co.; and C. A. Reynolds, 2004. How to Manage Pests. Weed Management in Lawns. Produced by University of California Statewide IPM Program. Published 1/04. http://www.ipm.ucdavis.edu/index.html

Levetin-McMahon, 2008. Plants and Society: origins of agriculture. Chapter11. Fifth Edition © The McGraw-Hill Companies,. Visit www.mhhe.com/levetin5e

Lewis M. J. T, 1994. "The Origins of the Wheelbarrow," *Technology and Culture*, Vol. 35, No. 3. (Jul., 1994), pp. 453–475 http://en.wikipedia.org/wiki/Wheelbarrow - cite_ref-27

LLC., 2012. *Story of farming*. History Source LLC. http://historylink101.com/lessons/farm-city/story-of-farming.htm

Lynn White, Jr., *Medieval Technology and Social Change* (Oxford: University Press, 1962), p. 42.

Machinery Manufacturers - Secondary Tillage. Worldwide Agricultural Machinery and Equipment Directory. 2006-11-14. Retrieved 2007-07-12.

Mano-E-Mano, 2010 The Most Influential People in the World: Nikolaus August Otto (1832-1891). http://www.blogger.com/profile/03508972403184413995

Marc Bloch, *French Rural History*, trans. Janet Sondheimer (Berkeley: University Press, 1966), 50.

Margaritis, Evi; Jones, Martin K.: "Greek and Roman Agriculture", in: Oleson, John Peter (ed.): *The Oxford Handbook of Engineering and Technology in the Classical World*, Oxford University Press, 2008, ISBN 978-0-19-518731-1, pp. 158–174 (166, 170)

Mary Bellis, 2014. The Agricultural Revolution. Introduction to the Agricultural Revolution. ©2014 About.com. http://inventors.about.com/

Mary Bellis, 2012. Antique tractors. http://inventors.about.com/od/articlesandresources/ig/Antique-Tractors---Photos.--0J/index.htm

Mary Bellis, 2012. History of the Railroad. http://inventors.about.com/bio/Mary-Bellis-496.htm

Mary Bellis, 2011. History of Farm Tractors - Minneapolis Steel and Machinery Company. Minneapolis, Minnesota (1902-1929) . http://inventors.about.com/bio/Mary-Bellis-496.htm

Mary Bellis, 2011.History of Tractors. http://inventors.about.com/bio/Mary-Bellis-496.htm

Mary Bellis, 2014. Machines to Cut Grains: Sickles & Reaper. ©2014 About.com. http://inventors.about.com/

McNeil, Ian (1990). *An Encyclopedia of the History of Technology*. London: Routledge. ISBN 0-415-14792-1

Mechanized agriculture From Wikipedia, the free encyclopedia. http://en.wikipedia.org/wiki/Productivity_improving_technologies_%28historical%29

Melander, B. 1997. Optimization of the adjustment of a vertical axis rotary brush weeder for intra-row weed control in row crops. *Journal of Agricultural Engineering Research* 68(1): 39-50.

Michael Hogan C., 2007. "Knossos fieldnotes", *The Modern Antiquarian*

Mifgs, 2006. The John Appleby Knotter. http://vitebskprint.com/m1a1-carc3/index.html

Mike Biscoe Land Clearing Hand Tools. © 1999-2014 Demand Media, Inc. http://www.ehow.com/

Mike Mann, 2013. Types of Land Clearing Equipment. © 1999-2014 Demand Media, Inc. http://www.ehow.com/. Last updated May 29, 2013

Mohd Taufik Bin Ahmad, 2012. Development of an Automated Mechanical Intra-Row Weeder for Vegetable Crops. *Graduate Theses and Dissertations*. Paper 12278.

Nancy M. Trautmann, Keith S. Porter and Robert J. Wagenet, 2012. Modern Agriculture: Its Effects on the Environment Natural Resources Cornell Cooperative Extension. http://psep.cce.cornell.edu/facts-slides-self/Factsheets.aspx

Neilson, N. (1936) Medieval Agrarian Economy. Henry Holt and Company.

Nicholas-Joseph Cugnot Biography (1725-1804). Copyright © 2014 Advameg, Inc. http://www.madehow.com/forum/

NIS, 1997. NIS Newsletter. Newsletter of the Nigerian Institutions of Surveyors. Vol.12 No. 2, 12pp. Victoria Island, Lagos.

Norman Einstein, 2012. *Map of the Fertile Crescent (Illustration) -- Ancient History Encyclopedia.htm published on 26 April 2012*

No Till Drills http://www.londononline.co.uk/factfile/historical/ population list on London online. Accessed 16 Jan 2014

Nwachukwu Chinweizu .N., 2006. Discuss in details the history of agriculture in Nigeria from the colonial era to the present day, pointing out clearly all agricultural programmes. http://www.onlinenigeria.com/ Posted to the web: 5/12/2006.

Olney, Ross. (1984) The Farm Combine. Walker Publishing Company.

Oliver Heritage, 2013. Oliver History: Hart-Parr. Copyright © 2013 Oliver Heritage. PO Box 519, Greenville, IL 62246-0519

Paul Hughes (3 March 2011). "Castlepollard venue to host Westmeath ing finals". *Westmeath Examiner*. Retrieved 01/06/2011.

Pete Daniel and Larry Jones, 1993. Preservation Strategies: Balancing Access, Use, Exhibition and Preservation Preserving the Hart-Parr Number 3. daniel_jones.pdf

Peter Bellwood, 2005. First Farmers: The Origins of Agricultural Societies. ISBN: 978-0-631205-66-1; 2005; 384 pages; Wiley-Blackwell, Malden MA; Phillip Edwards La Trobe University, VIC.http://rsj.e-contentmanagement.com/

Peter Maniate, 2011. Dog Power (Articles). http://hannibalkennels.on.ca/. Nov 5, 2011

Physical Weeding.2011.Physical weeding: Flame weeding: Machine design. Available at http://www.physicalweeding.com/flameweeding/index.html. Accessed 7 July 2011.Tornado.2011. Cultivation. Available at www.tornadosprayers.com.au/cultivation.htm. Accessed 25 June 2011.

Pfeiffer, J. E., 1977, The emergence of society: a prehistory of the establishment, McGraw Hill, New York.

Problems of Agricultural Mechanization. http://joeleebass.hubpages.com/ Last updated on July 28, 2012.

Robert Greenberger, *The Technology of Ancient China* (New York: Rosen Publishing Group, Inc., 2006), pp. 11-12.

Richard Trevithick 1771 - 1833 © 2003 - 2010 David Carvey/www.themagicofcornwall.com. http://www.themagicofcornwall.com/index.htm

Richard Van Vlecks, 200. Animal Treadmills on the Farm © 2002, American Artifacts, Taneytown Maryland.

Richard Van Vleck, 1998. Early Cow Milking Machines. © 1996, 1998, American Artifacts. Reprinted from SMMA issue 20.http://www.americanartifacts.com/smma/index.htm

Robert Brochier, 2006. John Froelich's tractor. Robert Brochier © 1997 - 2006. http://www.oocities.org/motorcity/downs/9828/fergy_accueil.htm

Rolt, L. T. C. & Allen, J. S., 1997. The steam engine of Thomas Newcomen, Ashbourne Landmark Publishing Ltd, pp 45.

Rumely, Edward A. (Edward Aloysius), 1882-1964. Guide to the Edward A. Rumely papers 1904-1959. http://nwda.orbiscascade.org/index.shtml

Sakrabani R., L. K. Deeks, M. G. Kibblewhite and K. Ritz, 2012. Impacts of Agriculture upon Soil Quality. ISBN: 978-1-84973-385-4 DOI:10.1039/9781849734974-00035

Sam Moore, 2003. The Journey of William Deering. http://www.farmcollector.com/ August 2003 Snell. Annals of the Labouring Poor. Ch. 4.

Sam Moore, 2008. Ten agricultural inventions that changed the face of farming in America. http://www.farmcollector.com/

Sarah Carter, *2012.* Agriculture and Agitation on the Oak River Dakota Reserve, 1875-1895. Published by the Manitoba Historical Society. *Page revised: 27 October 2012.* http://www.mhs.mb.ca/docs/mb_history/index.shtml

"Scottish Engineering Hall of Fame". *engineeringhalloffame.org.* 2012. Retrieved 27 August 2012.

Sommers, L. E., 1977. "Chemical composition of sewage sludge and analysis of their potential use as fertilizers". *J. Environ. Qual.* 6, 225–229.

Silver, 2012. Advantages of Modern Agriculture. *StudyMode.com.* Retrieved 12, 2012, from http://www.studymode.com/essays/Advantages-Of-Modern-Agriculture-1287077.html

Site selection and general planning. http://Downloads/x6708e02.htm

Smithsonian Magazine, 2007"The Mystery of Easter Island", April 01, 2007

Steam-Powered Car Invented by Nicolas Joseph Cugnot (1725-1804). ©AllPosters.com. Item #: 1345178. http://www.allposters.com/

Stephen Thompson, 2007.1860 Wagga Wagga Chinese Harrow. Museum of the Riverina, Wagga Wagga, Australia. June 2007. Crown ©2007. http://www.migrationheritage.nsw.gov.au/exhibition/objectsthroughtime-history/about-objects-through-time/

Steven , seed drill. http://irinventions.biss.wikispaces.net/Steven

The Howard Rotavator. 2007-02-03. Retrieved 2007-07-12.

The John Deere Tractor Legacy, 2003. Voyageur Press. 2003-10-30. ISBN 978-0-89658-619-2.

The Plough. http://www.machine-history.com/

The Rotherham tractor. www.rotherhamweb.co.uk/h/.htm

Thirsk. 'Walter Blith' in Oxford Dictionary of National Biography online edn, Jan 2008

"Tillage Equipment". Natural Resources Conservation Service. Retrieved 11 June 2012.

TimeMaps, 2014. The world of Hunter Gatherers. http://www.timemaps.com/home © 2014 TimeMaps Ltd.

Tjeerd H. van Andel, Eberhard Zangger, Anne Demitrack. "Land Use and Soil Erosion in Prehistoric and Historical Greece". *Journal of Field Archaeology*: 379–396. doi:10.1179/009346990791548628

Thomas Newcomen (1664-1729) and the atmospheric beam engine. Cornish Mining World Heritage. © Cornwall Council 2011

Tractors in the 1930s http://www.livinghistoryfarm.org/farminginthe30s/machines_04.html

Treadmillreview.net. The Treadmill: A History. http://www.treadmillreviews.net/treadmill-articles/. © *Copyright 2014 TreadmillReviews.net*

Trevor Watkins 2010. New light on Neolithic revolution in south-west Asia Antiquity, Vol.84 (2010) URL: http://antiquity.ac.uk/ant/084/0621/ant0840621.pdf

Tillett, N. D., Hague T., Grundy A. C., and Dedousis A. P., 2008. Mechanical within-row weed control for transplanted crops using computer vision. *Biosystems Engineering* 99 (2) (February): 171-178. doi:10.1016/j.biosystemseng.2007.09.026.

Tim Lambert, 2014. A brief history of inventions. http://www.localhistories.org/techtime.html

Tim Lambert, 2012. A brief history of the industrial revolution. http://www.localhistories.org/index.html

Upadhyaya, M. K., and R. E. Blackshaw. 2007. Non-chemical weed management: synopsis, integration and the future. In *Non-Chemical Weed Management*, Upadhyaya M.K and R E Blackshaw, ed.201-209. CAB International.

USDA. A History of American Agriculture 1776-1990. http://inventors.about.com/library/inventors/blfarm.htm

Wadley and Angus, 1993. The origins of agriculture: a biological perspective and a new hypothesis Published in *Australian Biologist* 6: 96-105, June 1993. www.ranprieur.com/readings/origins.html

Wang Zhongshu, trans. by K.C. Chang and Collaborators, *Han Civilization* (New Haven and London: Yale University Press, 1982).

Watt James von Breda. http://en.wikipedia.org/wiki/File_talk:Watt_James_von_Breda.jpg

Weide, R. Y. V. D., P. O. Bleeker, V. T. J. M. Achten, L. A. P. Lotz, F. Fogelberg, and B. Melander. 2008. Innovation in mechanical weed control in crop rows. *Weed Research* 48 (3): 215-224. http://dx.doi.org/10.1111/j.1365-3180.2008.00629.x.

White, *Medieval Technology*, p. 50

White, K. D. (1984): *Greek and Roman Technology*, London: Thames and Hudson, p. 59.

Wikipedia, 2014. Agricultural cooperative. http://en.wikipedia.org/wiki/Agricultural_cooperative

Wikipedia, 2014. Agricultural engineering. http://en.wikipedia.org/wiki/Agricultural_engineering

Wikipedia, 2014. Cultivator. http://en.wikipedia.org/wiki/Cultivator

Wikipedia, 2014. British Agricultural Revolution. http://en.wikipedia.org/wiki/Talk:British_Agricultural_Revolution

Wikipedia, 2014. Dibber. http://en.wikipedia.org/wiki/Dibber

Wikipedia, 2014. Edward P. Allis. http://en.wikipedia.org/wiki/Edward_P._Allis

Wikipedia, 2014. Green Revolution. http://en.wikipedia.org/wiki/Green_Revolution

Wikipedia, 2014. History of agriculture. http://en.wikipedia.org/wiki/History_of_agriculture

Wikipedia, 2014. Jean de Hautefeuille. http://en.wikipedia.org/wiki/Jean_de_Hautefeuille

Wikipedia, 2014. Leander J. McCormick. http://en.wikipedia.org/wiki/Leander_J._McCormick

Wikipedia, 2014. Marsh, Steward & Company. From Wikipedia, the free encyclopedia. url: http://en.wikipedia.org/wiki/File:Marsh, Steward & Company - Wikipedia, the free encyclopedia.htm

Wikipedia, 2014. Mechanized agriculturehttp://en.wikipedia.org/wiki/Mechanised_agriculture

Wikipedia, 2014. Nicolas-Joseph Cugnot. http://en.wikipedia.org/wiki/Nicolas-Joseph_Cugnot

Wikipedia, 2014. Nikolaus Ottohttp://en.wikipedia.org/wiki/Nikolaus_Otto

Wikipedia, 2014. Obed Hussey. http://en.wikipedia.org/wiki/Obed_Hussey

Wikipedia, 2014. Oliver Farm Equipment Company. http://en.wikipedia.org/wiki/Oliver_Farm_Equipment_Company

Wikipedia, 2014. Patrick Bell. http://en.wikipedia.org/wiki/Patrick_Bell

Wikipedia, 2014. Plough. http://en.wikipedia.org/wiki/Plough

Wikipedia, 2014. Self-propelled. http://en.wikipedia.org/wiki/Self-propelled

Wikipedia, 2014. W. W. Marsh. http://www.ancient.eu.com/http://en.wikipedia.org/wiki/File:Marsh_Harvester_1860 .html

Wikipedia, the free encyclopedia, 2013. British Agricultural Revolution http://en.wikipedia.org/wiki/File:Question_book-new.svg

Wikipedia, 2014. John Appleby (inventor). http://en.wikipedia.org/wiki/John_Appleby_%28inventor%29

Wikipedia, 2014. William Deering. http://en.wikipedia.org/wiki/William_Deering

William W. Bottorff,. What Was The First Car? A Quick History of the Automobile for Young People. www.ausbcomp.com/~bbott/cars/carhist.htm. Accessed April 14 2014.

Worthington, W., "50 Years of Agricultural Tractor Development," SAE Technical Paper 660584, 1966, doi:10.4271/660584.

Zenodot Verlagsgesellschaft mbH, 2012. Ploughing Egyptian Farmer. Ancient History Encyclopedia Xulon Press. Xulon Press. June 2002. ISBN 978-1-59160-134-0.

Appendix

A timeline of American agricultural development (1776-1990)

16th -18th Centuries	18th century - Oxen and horses for power, crude wooden ploughs, all sowing by hand, cultivating by hoe, hay and grain cutting with sickle, and threshing with flail
1776-99	1790's - Cradle and scythe introduced 1793 - Invention of cotton gin 1794 - Thomas Jefferson's mouldboard of least resistance tested 1797 - Charles Newbold patented first cast-iron plough

1800-1899

1800 1810 1820	1819 - Jethro Wood patented iron plough with interchangeable parts 1819-25 - U.S. food canning industry established
1830	1830 - About 250-300 labor-hours required to produce 100 bushels (5 acres) of wheat with walking plough, brush harrow, hand broadcast of seed, sickle, and flail 1834 - McCormick reaper patented 1834 - John Lane began to manufacture ploughs faced with steel saw blades 1837 - John Deere and Leonard Andrus began manufacturing steel ploughs 1837 - Practical threshing machine patented
1840	1840's - The growing use of factory-made agricultural machinery increased farmers' need for cash and encouraged commercial farming 1841 - Practical grain drill patented 1842 - First grain elevator, Buffalo, NY 1844 - Practical mowing machine patented 1847 - Irrigation begun in Utah 1849 - Mixed chemical fertilizers sold commercially
1850	1850 - About 75-90 labor-hours required to produce 100 bushels of corn (2-1/2 acres) with walking plough, harrow, and hand planting 1850-70 - Expanded market demand for agricultural products brought adoption of improved technology and resulting increases in farm production 1854 - Self-governing windmill perfected 1856 - 2-horse straddle-row cultivator patented

Timeline of agricultural mechanization

1860	1862-75 - Change from hand power to horses characterized the first American agricultural revolution 1865-75 - Gang ploughs and sulky ploughs came into use 1868 - Steam tractors were tried out 1869 - Spring-tooth harrow or seedbed preparation appeared

1870	1870's - Silos came into use
	1870's - Deep-well drilling first widely used
	    1874 - Glidden barbed wire patented
	    1874 - Availability of barbed wire allowed fencing of rangeland, ending era of unrestricted, open-range grazing

1880	1880 - William Deering put 3,000 twine binders on the market 1884-90 - Horse-drawn combine used in Pacific coast wheat areas
1890	1890-95 - Cream separators came into wide use 1890-99 - Average annual consumption of commercial fertilizer: 1,845,900 tons 1890's - Agriculture became increasingly mechanized and commercialized 1890 - 35-40 labor-hours required to produce 100 bushels (2-1/2 acres) of corn with 2-bottom gang plough, disk and peg-tooth harrow, and 2-row planter 1890 - 40-50 labor-hours required to produce 100 bushels (5 acres) of wheat with gang plough, seeder, harrow, binder, thresher, wagons, and horses 1890 - Most basic potentialities of agricultural machinery that was dependent on horsepower had been discovered

1900-2000

1900	1900-1909 - Average annual consumption of commercial fertilizer: 3,738,300 1900-1910 - George Washington Carver, director of agricultural research at Tuskegee Institute, pioneered in finding new uses for peanuts, sweet potatoes, and soybeans, thus helping to diversify southern agriculture.
1910	1910-15 - Big open-geared gas tractors came into use in areas of extensive farming 1910-19 - Average annual consumption of commercial fertilizer: 6,116,700 tons 1915-20 - Enclosed gears developed for tractor 1918 - Small prairie-type combine with auxiliary engine introduced
1920	1920-29 - Average annual consumption of commercial fertilizer: 6,845,800 tons

	1920-40 - Gradual increase in farm production resulted from expanded use of mechanized power 1926 - Cotton-stripper developed for High Plains 1926 - Successful light tractor developed
1930	1930-39 - Average annual consumption of commercial fertilizer: 6,599,913 tons 1930's - All-purpose, rubber-tired tractor with complementary machinery came into wide use 1930 - One farmer supplied 9.8 persons in the United States and abroad 1930 - 15-20 labor-hours required to produce 100 bushels (2-1/2 acres) of corn with 2-bottom gang plough, 7-foot tandem disk, 4-section harrow, and 2-row planters, cultivators, and pickers 1930 - 15-20 labor-hours required to produce 100 bushels (5 acres) of wheat with 3-bottom gang plough, tractor, 10-foot tandem disk, harrow, 12-foot combine, and trucks
1940	1940-49 - Average annual consumption of commercial fertilizer: 13,590,466 tons 1940 - One farmer supplied 10.7 persons in the United States and abroad 1941-45 - Frozen foods popularized 1942 - Spindle cotton picker produced commercially 1945-70 - Change from horses to tractors and the adoption of a group of technological practices characterized the second American agriculture agricultural revolution 1945 - 10-14 labor-hours required to produce 100 bushels (2 acres) of corn with tractor, 3-bottom plough, 10-foot tandem disk, 4-section harrow, 4-row planters and cultivators, and 2-row picker 1945 - 42 labor-hours required to produce 100 pounds (2/5 acre) of lint cotton with 2 mules, 1-row plough, 1-row cultivator, hand how, and hand pick
1950	1950-59 - Average annual consumption of commercial fertilizer: 22,340,666 tons 1950 - One farmer supplied 15.5 persons in the United States and abroad 1954 - Number of tractors on farms exceeded the number of horses and mules for first times 1955 - 6-12 labor-hours required to produce 100 bushels (4 acres) of wheat with tractor, 10-foot plough, 12-foot role weeder, harrow, 14-foot drill and self-propelled combine, and trucks Late 1950's - 1960's - Anhydrous ammonia increasingly used as cheap source of nitrogen, spurring higher yields
1960	1960-69 - Average annual consumption of commercial fertilizer: 32,373,713

	tons
	1960 - One farmer supplied 25.8 persons in the United States and abroad
	1965 - 5 labor-hours required to produce 100 pounds (1/5 acre) of lint cotton with tractor, 2-row stalk cutter, 14-foot disk, 4-row bedder, planter, and cultivator, and 2-row harvester
	1965 - 5 labor-hours required to produce 100 bushels (3 1/3 acres) of wheat with tractor, 12-foot plough, 14-foot drill, 14-foot self-propelled combine, and trucks
	1965 - 99% of sugar beets harvested mechanically
	1965 - Federal loans and grants for water/sewer systems began
	1968 - 96% of cotton harvested mechanically
1970	1970's - No-tillage agriculture popularized
	1970 - One farmer supplied 75.8 persons in the United States and abroad
	1975 - 2-3 labor-hours required to produce 100 pounds (1/5 acre) of lint cotton with tractor, 2-row stalk cutter, 20-foot disk, 4-row bedder and planter, 4-row cultivator with herbicide applicator, and 2-row harvester
	1975 - 3-3/4 labor-hours required to produce 100 bushels (3 acres) of wheat with tractor, 30-foot sweep disk, 27-foot drill, 22-foot self-propelled combine, and trucks
	1975 - 3-1/3 labor-hours required to produce 100 bushels (1-1/8 acres) of corn with tractor, 5-bottom plough, 20-foot tandem disk, planter, 20-foot herbicide applicator, 12-foot self-propelled combine, and trucks
1980-90	1980's - More farmers used no-till or low-till methods to curb erosion
	1987 - 1-1/2 to 2 labor-hours required to produce 100 pounds (1/5 acre) of lint cotton with tractor, 4-row stalk cutter, 20-foot disk, 6-row bedder and planter, 6-row cultivator with herbicide applicator, and 4-row harvester
	1987 - 3 labor-hours required to produce 100 bushels (3 acres) of wheat with tractor, 35-foot sweep disk, 30-foot drill, 25-foot self-propelled combine, and trucks
	1987 - 2-3/4 labor-hours required to produce 100 bushels (1-1/8 acres) of corn with tractor, 5-bottom plough, 25-foot tandem disk, planter, 25-foot herbicide applicator, 15-foot self-propelled combine, and trucks
	1989 - After several slow years, the sale of farm equipment rebounded
	1989 - More farmers began to use low-input sustainable agriculture (LISA) techniques to decrease chemical applications

Source: http://inventors.about.com/library/inventors/blfarm1.htm

Titles in author's publication series

Agriculture & mechanization series

- ♣ Farm power and machinery operations
- ♣ Agricultural machinery & mechanization
- ♣ Agricultural engineering: principles and practice (Vol 1)
- ♣ Agricultural engineering: principles and practice (Vol 2)
- ♣ Farm tractor systems: operations and maintenance
- ♣ Timeline of agricultural mechanization

Horticultural series

- ♣ Horticultural machinery: equipment and safety
- ♣ Fruits and vegetable technologies: management options

Workplace safety and machine technology series

- ♣ Agricultural machinery hazards & safety practices
- ♣ Workplace hazards risks & control
- ♣ Workshop technology & practice
- ♣ Technical drawing presentation and practice

Sustainable agriculture and environment series

- ♣ Sustainable agriculture: prospects and challenges
- ♣ Sustainable environmental management: issues and projections

Students' handbook series

- ♣ Study companion
- ♣ Path to exam success

Copies available @ http://www.amazon.com/Segun-R.-Bello/e/B008AL6RI0

www.ingramcontent.com/pod-product-compliance
Lightning Source LLC
Chambersburg PA
CBHW081718170526
45167CB00009B/3614